Building a Travel Risk Management Program

Building a Travel Risk Management Program

Traveler Safety and Duty of Care for Any Organization

Charles Brossman

AMSTERDAM • BOSTON • HEIDELBERG • LONDON
NEW YORK • OXFORD • PARIS • SAN DIEGO
SAN FRANCISCO • SINGAPORE • SYDNEY • TOKYO
Butterworth-Heinemann is an imprint of Elsevier

British Library Cataloguing-in-Publication Data
A catalogue record for this book is available from the British Library.

Library of Congress Cataloging-in-Publication Data
A catalog record for this book is available from the Library of Congress.

ISBN: 978-0-12-801925-2

For Information on all Butterworth-Heinemann publications
visit our website at https://www.elsevier.com/

Working together
to grow libraries in
developing countries

ELSEVIER   Book Aid International

www.elsevier.com • www.bookaid.org

Publisher: Candice Janco
Acquisition Editor: Sara Scott
Editorial Project Manager: Hilary Carr
Production Project Manager: Punithavathy Govindaradjane
Designer: Mark Rogers

Typeset by MPS Limited, Chennai, India

Contents

Foreword

You likely picked up this book for one of a few reasons. You may be asking "What is travel risk management or TRM?" Or, you may have heard of TRM in passing or through your work in the field of security, risk, travel, hospitality, human resources, insurance, business continuity, or emergency/crisis management, or possibly as a manager sending people out into the world. Even more likely, and without even knowing it, you have been one of the hundreds of millions of travelers supported by a TRM program. As a professional, or even a casual traveler flipping through the book, you should find the topics interesting and relevant. TRM is all about minimizing the risks or hazards that people face when then are traveling and helping to ensure that they return home safely and have a productive (and fun) trip.

I left the U.S. Intelligence Community in the spring of 1999, having worked with several of the three-letter agencies over a 17-year career. I had the honor of working with amazing people, many of whom put their lives on the line every day. Over this time, I worked on intelligence systems that supported both our national decision makers and the warfighter in the field and in battle. This was my background and experience when I became one of the founders, and ultimately the CEO, of iJET International in the fall of 1999. iJET coined the term *travel risk management* and we have been the leader in both defining and advancing the application of risk management principles to travel. I have the opportunity to write, speak, and teach on the topic of TRM throughout the world. I am proud and humbled when my colleagues refer to me as the "father of TRM."

I met Charles in 2004, when he was working for a travel-tracking technology company called FlightLock. Over the years, Charles has dedicated himself to learning, applying, and developing TRM solutions for the travel industry. We have spent many hours discussing a wide range of TRM issues. Recently, Charles has been one of our key contributors in developing the next evolution of the Travel Risk Management Maturity Model (TRM3), a tool to enable organizations to assess and benchmark their TRM program. In this book, Charles has pulled together a wealth of information not previously compiled and organized for the student or practitioner. He guides you through the various elements of TRM and provides his insight and experience along the way.

As with most things in life—from driving a car to cooking—they seem simple on the surface but are quite complex as you delve into the details. This is true for TRM as well. Applying risk management principles and processes to travel is where it all starts. However, just defining travel and all of its components can be challenging as it covers the entire end-to-end experience of the journey. When discussing the processes related to the planning, purchasing, provisioning, and delivering of travel-related services, most organizations use the term *travel management* or *journey management*. In the book, Charles focuses on air, rail, hotels, and ground transportation as the key components of travel. He shows that even ground transportation is a complex component consisting of personal automobiles, car rentals, limousine companies, and even ride-sharing services. I was pleased to see that he devoted a chapter to travel management companies, as they play a critical role in both implementing and sustaining a TRM program, especially when it comes to providing high-quality travel data and supporting travelers through their journey.

A key challenge faced by practitioners is to find the resources and money to implement and support a robust TRM program. It is often difficult to quantify a return-on-investment (ROI) when you are in the problem avoidance business. How do you quantify the amount of money that wasn't spent when a traveler doesn't get sick or kidnapped or in an automobile accident? Charles takes this challenge head-on in the chapter "Finding the Money for Travel Risk Management."

As with both Charles and myself, I hope that you get the TRM bug! Getting the TRM bug is going beyond just learning about the discipline; it is being drawn into the cause by doing your part in helping to keep your clients, friends, loved ones—even yourself—safe and healthy when traveling.

D. Bruce McIndoe
Chairman & CEO, iJET International, Inc.
Washington, D.C. USA

Introduction

Before I dive into how this book can help you, or what to be mindful of as you read it, I want to start this journey with you based upon what we can, hopefully, all agree. The most important asset of all is the human life. Not all of us know what it feels like to feel helpless in a situation where the life of someone dear to us is in jeopardy, yet many of us can relate to the memories of watching various acts of terrorism unfold via the media over recent years. While bad things are still going to happen, and not all of them can be prevented, both employers and travelers who learn from travel risk management (TRM) concepts can mitigate or prevent incidents far better than if they have no plans, strategy, or training to protect that which should be valued more than anything. No one wants to feel helpless in the event of a crisis.

One of my biggest goals for this book is to change corporate perception of TRM, which is sometimes minimized by denial or ineffective "shortcuts" that fall short of industry standards, putting minimal savings above the safety of travelers. With various other areas of risk that companies invest in and legitimize on a large scale, such as physical security (e.g., facilities) or financial risks, often not nearly as much thought goes into how safe a traveler or expatriate's experience will be while traveling abroad, representing their employer.

Ask yourself:

- Does any thought go into the safety of the hotels that you choose as preferred suppliers, and to what extent?
- What about training? Or the safety of air or ground transportation?
- What would you tell a spouse of someone who was kidnapped while traveling on business for your company, if you had no strategic kidnap and ransom plans or protocols?

What often happens is that travel security or TRM is delegated to physical security executives or resources dedicated to other aspects of traditional security, when they are unfamiliar with the intricacies of travel as a spend category and all of the specialized training and understanding that is required to be effective in this area. Just as travel is unique to manage from a procurement perspective (versus other spend categories like office supplies or raw materials), it is also very unique from a risk perspective. Travel managers aren't typically equipped to address or manage TRM alone (particularly those without a travel industry background). However, many are uniquely positioned, along with limited and specific travel management company (TMC) or travel agency support, to collaborate with TRM specialists, internal security departments, and other departmental stakeholders, to create and manage productive programs. In general, TRM should always involve a cross-section of company stakeholders, but TRM most definitely needs someone with a deep understanding of travel data, TMC operations, travel technology, and an aptitude for learning and applying changing TRM best practices over time.

In this text, you will learn about industry standards for a TRM program framework and metrics, along with case studies and best practices. However, most importantly, you will learn that each employer's approach to TRM is uniquely different, like a recipe, even when following the standard principles provided. This is because of each company's distinct culture, risk tolerance, industry, geographic location, and many other factors. Although each recipe will vary, each should follow the same framework and year-over-year continuous process improvement approach. No shortcuts.

As our world changes, we adapt to new or modified business models with traditional suppliers such as airlines, while we grapple with the impact of new entrants in the travel market, such as sharing economy suppliers. Additionally, we learn from tragedies, such as natural disasters and other factors, over time, and try to share this learning as a community for the greater good. As travel changes, so must how we manage the risks involved with it.

I hope that you benefit from "Building a Travel Risk Management Program" and pass along these approaches and best practices to your friends and colleagues, making traveling for business safer for all of us.

Charles Brossman
Travel Risk Management Consultant
Email: Charles@charlesbrossman.com
Twitter: @travelcharles
LinkedIn: http://www.linkedin.com/in/brossman
Press kit: https://www.presskit.to/charlesbrossman
Web: http://www.charlesbrossman.com

About the Author

Mr. Brossman is an internationally recognized expert, speaker and writer on travel risk management. He is a former corporate travel manager, and has held senior level positions at global travel management companies as the sole travel risk management subject matter expert covering over 150 countries, specializing in developing and implementing travel risk management products and services around the world. Mr. Brossman is a former member of the GBTA Risk Committee and the GBTA Foundation Risk Task Force, and currently sits on the advisory board for the Global Congress on Travel Risk Management, influencing industry best practices and teaching them to corporate clients and organization members at conferences, meetings and webinars throughout the year. Learn more about Charles at charlesbrossman.com, and follow him on Twitter at @travelcharles. His email address is Charles@charlesbrossman.com. A presskit for Mr. Brossman can be found at https://www.presskit.to/charlesbrossman.

Planning for known and unknown risks

Prior to diving into the various aspects of building a travel risk management (TRM) program in the subsequent chapters, the purpose of this chapter will be to broaden your perception of why each and every company must address TRM at some level. The chapter begins with the corporate obligation of "duty of care" and what that means at a fundamental level, and then provides examples of different kinds of risks that companies should think about and implement plans to address. There are an infinite number of potential use cases for risk exposure to travelers, but these examples provide good food for thought, in particular to those companies whose knee-jerk reactions to creating a TRM program is typically comments about their not necessarily needing one because they don't believe that they travel to high-risk destinations, which is a farce.

As you will learn throughout this text, risk exposure is not always directly related to the risk rating of a particular destination as provided by risk intelligence providers. It can also be about risks that are specific to a traveler, their behavior and any number of other factors, some of which may be foreseeable, and some not. This information is important, but in the absence of a moderate to high risk rating, there is still the potential for an individual or widespread crisis that can affect groups of people and even an entire company. Subsequent chapters will delve into greater detail on some more common risk factors, along with case studies and precedents.

Legal duty of care—definition[1]

"Duty of care" stands for the principle that directors and officers of a corporation in making all decisions in their capacities as corporate fiduciaries, must act in the same manner as would a reasonably prudent person in their position.

Courts will generally adjudge lawsuits against director and officer actions to meet the duty of care, under the business judgment rule. **The business judgment rule stands for the principle that courts will not second guess the business judgment of corporate managers and will find the duty of care has been met so long as the fiduciary executed a reasonably informed, good faith, rational judgment without the presence of a conflict of interest.** The burden of proof lies with the plaintiff to prove that this standard has not been met. If the plaintiff meets the burden, the defendant fiduciary can still meet the duty of care by showing entire fairness, meaning that both a fair process was used to reach the decision and that the decision produced a substantively fair outcome for the corporation's shareholders.

[1] Cornell University Law School, "Duty of Care: Definition," http://www.law.cornell.edu/wex/duty_of_care.

iJET International defines "Duty of Care" specific to TRM as follows:[2]

Duty of Care: *This is the legal responsibility of an organization to do everything "reasonably practical" to protect the health and safety of employees. Though interpretation of this language will likely vary with the degree of risk, this obligation exposes an organization to liability if a traveler suffers harm. Some of the specific elements encompassed by Duty of Care include:*

- *A safe working environment—this extends to hotels, airlines, rental cars, etc.*
- *Providing information and instruction on potential hazards and supervision in safe work (in this case, travel)*
- *Monitoring the health and safety of employees and keeping good records*
- *Employment of qualified persons to provide health and safety advice*
- *Monitoring conditions at any workplace (including remote locations) under the organization's control and management*

Relative to "Duty of Care" is the "Standard of Care" that companies are compared to in defending what is "reasonable best efforts" or "reasonably practical," based upon what resources and programs are put into place by an organization's peers to keep travelers safe.

Prior to 2001, business travelers thought nothing of being able to walk into an airport and meet their loved ones at their arrival gate. No security barriers, no cause for concern because air travel was something that at the time, our collective psyche felt generally safe, with the exception of a hijacking upon occasion. Fast forward to a post-9/11 world, and consider what the world's airports look like now and how the processes surrounding airport security have changed the way that we travel, whether for business or pleasure.

Why would any of us believe that the need for added security, particularly around those traveling for business, begins and ends at the airport? For companies who have been paying attention since 9/11, the ones who, outside of the public eye, have had to deal with critical incidents that had the potential for loss of lives, corporate liability, and damage to their company's reputation, having a structured TRM program not only reduced the potential for risk, but heightened the awareness of risk to their travelers. Their definition of "travelers" extended beyond employees (transient travelers to expatriates) to contractors, subcontractors, and dependents. Keeping travelers aware of imminent dangers takes effort and planning, and isn't something that employers can any longer react to after the fact. In some countries, lack of planning or resources to support business travelers has the potential to be grounds for claims of negligence in a company's duty of care responsibilities, and can lead to a criminal offense, such as with the United Kingdom's (UK) Corporate Manslaughter and Corporate Homicide Act of 2007. What the "business judgment rule" in the above duty of care definition means in layman's terms is that a company must be able to prove that it put forth reasonable best efforts to keep its travelers safe. How this applies in different circumstances, jurisdictions and countries will vary. Most countries' duty of care requirements fall under their occupational safety and health laws. **For a comprehensive list**

[2] iJET, "White Papers: Duty of Care," http://info.ijet.com/resources/whitepaper.

of occupational health and safety legislation by country, an updated global database is maintained by the International Labour Organization (www.ilo.org[3]).
Simply put, companies cannot afford to no longer have a proactive TRM program
and just react after an incident takes place. The end result could reflect negligence
on behalf of the company. For extensive detail on the UK's definition of duty of care
in relation to the Corporate Manslaughter and Corporate Homicide Act of 2007, visit
http://www.legislation.gov.uk/ukpga/2007/19.

Duty of care and tort law in the United States

Because each of the 50 U.S. states is a separate sovereign free to develop its own tort
law under the Tenth Amendment, there are several tests to consider for finding a duty
of care under U.S. tort law, in the absence of a federal law.

Tests include:

- Foreseeability—In some states, the only test is whether the harm to the plaintiff that resulted
  from the defendant's actions was foreseeable.
- Multifactor test—California has developed a complex balancing test consisting of multiple
  factors that must be carefully weighed against one another to determine whether a duty of
  care exists in a negligence action.

California Civil Code section 1714 imposes a general duty of ordinary care, which
by default requires all persons to take "reasonable measures" to prevent harm to
others. In the 1968 case of Rowland v. Christian (**after and based on this case, the
majority of states adopted this or similar standards**), the court held that judicial
exceptions to this general duty of care should only be created if clearly justified based
on the following public-policy factors:

- The foreseeability of harm to the injured party;
- The degree of certainty that he or she suffered injury;
- The closeness of the connection between the defendant's conduct and the injury suffered;
- The moral blame attached to the defendant's conduct;
- The policy of preventing future harm;
- The extent of the burden to the defendant and the consequences to the community of impos-
  ing a duty of care with resulting liability for breach; and the availability, cost, and preva-
  lence of insurance for the risk involved;
- The social utility of the defendant's conduct from which the injury arose.

**A 2011 law review article identified 43 states that use a multifactor analysis in
23 various incarnations and consolidated them into a list of 42 different factors
used by U.S. courts to determine whether a duty of care exists.**

Pioneering companies (often in the energy services sector or government contractors) who were some of the first to adopt and implement forward-thinking programs,
recognized early on that a critical incident or "crisis," isn't usually defined as an
event impacting large numbers of people. They found that the largest percentages

[3] International Labour Organization, http://www.ilo.org/dyn/legosh/en/f?p=14100:1000:31633078050819
::NO:::.

of incidents that required support, involved individual travelers or small groups. So while policies, plans, and readiness exercises are good to have in place for those highly visible incidents impacting large numbers of people, if handled improperly, the smaller incidents can cost companies considerably in damages and litigation costs, should their travelers or their travelers' surviving families prove that the companies in question weren't properly prepared to handle such incidents as they arise.

Case Study—U.S. Workers Compensation and Arbitration

Khan v. Parsons Global Services, Ltd

United States Court of Appeals, District of Columbia Circuit—Decided April 11, 2008 (https://www.cadc.uscourts.gov/internet/opinions.nsf/8DD6474D9DD 96BCE85257800004F879D/$file/07-7059-1110404.pdf)

- During the course of employment in the Philippines, on a day off, Mr. Khan was kidnapped and subsequently tortured.
- Employment contract included a broadly worded arbitration clause, and a separate clause specifying "workers compensation insurance" as "full and exclusive compensation for any compensable bodily injury" should damages be sought.
- Allegations that employer's disregard for Mr. Khan's safety in favor of minimizing future corporate kidnappings considering the way Parsons handled the situation provoked Mr. Khan's kidnappers to torture him, cutting of a piece of his ear, sending a video tape of the incident to the employer, causing the Khans severe mental distress.
- Mrs. Khan alleged efforts by the employer to prevent her from privately paying the ransom, despite threats of torture, may have exposed Mrs. Khan to guilt of knowing that she could have prevented Mr. Khan's suffering if the employer had not withheld the ransom details from her.
- Mr. and Mrs. Khan filed a lawsuit for Parsons' alleged mishandling of ransom demands by the kidnappers, and also alleging negligence and intentional infliction of emotional distress in D.C. Superior Court in 2003.
- The employer removed the case to the federal district court, arguing on the merits of the New York Convention for the Recognition and Enforcement of Foreign Arbitral Awards, and then filed a single motion to dismiss or, as an alternative, to obtain summary judgment to compel arbitration.
- The employer initially received a summary judgment to compel arbitration.
- Upon appeal, this judgment was reversed. The court found that the recovery of the Khans' tort claims were not limited by Mr. Khan's contract to workers' compensation insurance.
- An additional appeal contended that the initial summary judgment granted by the court denied the Khan's discovery requests, and dismissed Mrs. Khan's claim for intentional infliction of emotional distress
- Through the appeals process, the court found that the employer had in effect waived their right to arbitration.

This case study calls into question legal jurisdiction, U.S. workers' compensation liability limitations for employers, and the value of being prepared for such an incident as kidnapping.

This chapter outlines at a high level general categories that all companies must take into consideration when developing a TRM program. Very often the question is asked, "Do I really need to do any of this, because our company hasn't been sued to date?" If you have employees or contractors traveling on your behalf (especially internationally), whereby your company is paying for their time and/or expenses, then the answer is absolutely yes. The level of investment and complexity may vary between companies, but in general, all companies must have a plan for how to address the issues provided herein and others. Duty of care is never finite in its definition because companies must consider how laws from one country to the next will apply to travelers, contractors, potential subcontractors, and expatriates and their dependents, as well as any potential for conflict of law. Also, as shown in the Khan v. Parsons Global Services, Ltd. case study listed earlier in this chapter, employer remedies such as worker's compensation insurance in the U.S. aren't absolute; and therefore, warrants additional efforts and protections. Consider the following incident types or risk exposures, which in some instances can impact large numbers of travelers, but more commonly impact only one person.

Examples of potential risk exposures and incident types

Medical issues or concerns

According to the U.S. Department of Commerce International Trade Administration, only 10 percent of international business travelers receive pretravel health care. Pretravel health care can include, but is not limited to things like new or updates to vaccinations or inoculations, general health exams, medical treatment or procedures for a condition that may be risky to travel with, or prescription medicine planning for travel lasting for extended periods (longer than 30 days).

The chief operating officer at iJET, John Rose, comments that, "A percentage of calls into our crisis response center are for minor, individual medical issues."

However, callers may not always know that the situation is minor until they reach someone for support, which is why having an easy-to-identify, easy-to-access, single contact number or hotline for medical and security support is so important to all companies. A contracted crisis support service will know based upon predetermined protocols, which providers will support the traveler in the part of the world where they are traveling for medical issues, and ensure that the traveler gets the immediate advice that they need from a vetted medical professional. Sometimes with a brief conversation with a nurse, the parties can determine a minor treatment that the traveler can facilitate, and in other circumstances a referral to a more senior medical official or emergency medical resource may be necessary based upon the initial consultation by the first-level medical support personnel contracted by the traveler's company. As discussed later in the book, who provides the crisis response case management and who provides the medical or security services specific to the traveler in question are not necessarily mutually exclusive. There could be different providers in different parts of the world, used for different reasons that are outlined in company policies and protocols.

The consequences of mistakes as a result of a lack of preparation or resources can be costly, from financial loss and traveler productivity loss to the company, to a serious health issue for the traveler, or simply a ruined trip.

While clarity via training and policies on who supports traveler medical issues should be very clear to everyone within an organization, the following common medical mistakes should be avoided where possible, as recommended by Dr. Sarah Kohl, MD of TravelReadyMD (http://www.TravelReadyMD.com):

Mistake 1: Assumption that vaccines are complete preparation for an overseas trip

Statistically, most medical problems you are likely to experience while traveling overseas cannot be prevented with a vaccine. For example, there are no vaccines for jet lag, diarrhea, blood clots, malaria, or viral infections such as dengue. Before you travel overseas, make sure you are educated about these potential problems. Most can be prevented with simple measures.

Mistake 2: Conflicting Internet information

Information from different sources on the Internet can be conflicting and can lead you to believe you need more interventions than actually necessary. As travelers prepare to depart, employers should provide them with access to resources that can advise on medical concerns relative to your destinations. Of course, travelers should also discuss any personal medical condition concerns with their own or qualified medical professionals in addition to receiving employer provided risk intelligence regarding their trip.

Mistake 3: Failing to make simple preparations for predictable health issues

Unfortunately, travelers regularly suffer needless medical complications because they fail to take simple steps to avoid predictable issues. Simple precautions can save you a lot of discomfort and make your trip safer and more enjoyable.

Here are some examples: medical compression stockings, if properly fitted, can protect you from a life-threatening blood clot. Knowing the right insect spray to choose, from the multitude of choices available, can protect you from insect-borne disease. Avoiding seemingly harmless activities in certain locations (ones that a hotel concierge might even recommend) can protect you from parasites, respiratory illness or malaria.

Mistake 4: Assuming the quality of care for chronic conditions abroad

Travelers often fail to recognize how a common illness such as diarrhea or a respiratory infection can cause a flare-up of an underlying condition. Travelers who are good at managing food allergies, asthma, and diabetes at home may experience difficulty finding the resources they need overseas. In addition, these individuals may find themselves looking to a non–English-speaking doctor for help.

Mistake 5: Assuming that travel to a Western-style country is travel to a low-risk country

Measles, tuberculosis, and other infections are gaining a foothold in some European countries. Low immunization rates within these communities are thought to be the root cause. Don't risk becoming ill or bringing an infection home. Check with your health care provider before you travel to discuss preventive measures.

If you have a chronic health problem that is well under control, you will want to be prepared to self-treat under certain conditions. You may also want to be prepared to access a network of doctors who speak your native language, if needed.

Lastly, travelers should never assume that a pre-existing condition is covered by corporate- or consumer-based travel insurance or medical membership programs. When in doubt, always ask your human resources department or TRM program administrator. Companies commonly expect that corporate insurance policies or Business Travel Accident (BTA) policies provide enough coverage for travelers, when sometimes they may not. This is why protocols and regular training exercises for internal risk program stakeholders take place, to understand what is covered and what is not, as well as how to handle each situation.

Whether insured or not, consider the value and cost savings of prevention based treatment as shown in the examples provided below.

Estimated examples of costs for prevention versus treatment of some potential medical issues while traveling abroad

Incident	Average prevention cost	Average cost of treatment	Average workdays lost
Malaria	US$162	US$25,000	6–24
Hepatitis A	US$200–$300	US$1800–$2500	27
Medical emergencies abroad	US$15–$370—cost of medical evacuation and support insurance for a 2-week trip	US$25,000–$250,000— cost of medical emergency without coverage	

Source: U.S. Department of Commerce International Trade Administration, "Business Pulse: Travelers' Health," http://www.cdcfoundation.org/businesspulse/travelers-health-infographic.

Biohazards, toxicity, epidemics, and pandemics

Consider the possibility that anything that an employee or representative comes in contact with during the course of a business trip (during or after hours) that can potentially make them ill or kill them is a liability to the employer.

Biological hazards or biohazards are pathogens that pose a threat to the health of a living organism, which can include medical waste, microorganisms, viruses, or toxins. Toxicity is the degree to which a substance can damage an organism (not exclusively biological, as it could be chemical).

Brett Vollus, a former Qantas airline employee of 27 years, filed suit against the airline claiming that his spraying of government-mandated insecticides on planes to prevent the spread of insect-related diseases like malaria, caused him to develop Parkinson disease after 17 years of administering the chemicals in the flight cabins. It was also discovered from a brain scan after a tripping incident that Vollus had a malignant brain tumor. Considering this was a government mandate, it will be interesting to see if the question becomes: What did the government know about the risks of these chemicals? If a precedent is set in this suit, will liability extend to other airlines using or who have used such chemicals for extended periods, against repeat business travelers who regularly flew or fly in markets where such spraying was or is common practice over a long period of time?

Epidemics are outbreaks of disease that far exceed expected population exposures during a defined period of time. Epidemics are usually restricted to a specific area, as opposed to pandemics that cover multiple countries or continents.

Mature TRM programs monitor these more visible outbreaks and recommend vaccinations for travelers going to impacted areas; they also provide access to emergency medical resources when necessary, but also have a large focus on education, training, and prevention. However, employers should always be mindful of other environmental factors in the traveler's workplace both at home or abroad, such as urban or rural environmental factors. Examples may include prolonged exposure to pollution, lack of sanitation (particularly when it comes to their expat communities). Employers should work towards limiting those exposures or changing the environment through continuous process improvement reviews.

According to major medical and security evacuations suppliers, corporate-sponsored evacuations involving one or more travelers happen almost every day when you include both medical and security-related evacuations. It is a mistake to think that just because a case study or example is slightly dated, the instances they represent occur infrequently. It's quite the opposite. However, most incidents are not publicly documented to the degree that they can be reported upon.

Pandemics

The five primary things that companies must be concerned with when facing a pandemic situation are:

1. The potential impact on personnel.
2. The pandemic, crisis response plan.
3. The potential impact on business operations.
4. The potential impact on business supply chain.
5. The potential impact to share value or price.

What many companies don't consider is the potential for shareholder lawsuits against executives for business losses resulting from a lack of planning for situations such as pandemics. From shared sick time policies to work-at-home policies during a crisis, being able to quickly communicate a position or a plan, and to answer questions in the event of such an emergency, can not only save money and productivity, but garner employee confidence and calm nerves. Chapter 9 elaborates on the relationship between travel risk management (TRM) and other aspects of risk management across the enterprise (ERM–enterprise risk management).

Ebola's impact on Fonterra's bottom line

According to the *New Zealand Herald*,[4] the country's largest company, Fonterra, could lose $150 million because of the Ebola epidemic. Fonterra CEO, Theo Spierings, noted that when African countries lock down their borders to control the disease, demand dropped for Fonterra's products. He commented, "So…movements in West Africa become more and more difficult, so that limits movement of food as well, movement of people—people going to the market, doing their groceries—so you see demand really dropping pretty fast." "If the market in West Africa slowed down or dropped off that would affect 100,000 tonnes of powder," Mr. Spierings said. "That's about 5 percent, 6 percent of our exports. So you talk…$150 million or something like that."

White Paper

Is your organization pandemic ready?

Prepared by
Reputation Management Services Group
Eric Mower and Associates Public Relations and Public Affairs
November 2009
For further information:
Peter Kapcio, pkapcio@mower.com
Evan Bloom, ebloom@mower.com
(315) 466–1000

Introduction

Most businesses have already begun to feel the impact of the H1N1 virus, with absenteeism rising.

Harvard's School of Public Health recently released survey data showing how deeply concerned U.S. businesses are about the possibility of widespread employee absenteeism that might follow an outbreak of the swine flu (H1N1).

Researchers from the school questioned more than 1000 businesses across the country. Two-thirds of companies said they couldn't operate normally if more than half of their workers were out for 2 weeks. And four of five organizations predicted severe operating problems if half of their workers missed a month of work.

---

[4] *New Zealand Herald*, "Ebola Epidemic Could Cost Fonterra $150m," October 25, 2014, http://www.nzherald.co.nz/business/news/article.cfm?c_id=3&objectid=11348296.

These survey results should encourage all organizations to prepare for the worst by developing a crisis management plan. In addition to ample warning, senior management has ample *reason* to prepare, and no excuse not to. An organization's executives won't be blamed for the outbreak, but they do risk censure if they fail to prepare, respond, and communicate with internal and external stakeholders.

This white paper tells how.

To help organizations and their leaders prepare for a possible H1N1 pandemic, certain key issues must be addressed to keep operations running as smoothly as possible:

• Human resource (HR) issues that drive pandemic planning.
• Planning for steps necessary to keep an organization operating during the pandemic period.
• Implementing steps needed to create an enterprise-wide crisis management plan.
• Internal and external issues that crisis communications must address.

Why bother planning for the H1N1 pandemic? To put it simply, companies and organizations that plan for any type of crisis demonstrate the behavior of responsible citizens. Formulating a detailed crisis management plan specifically for H1N1 achieves four things:

1. Protects employees' health and safety.
2. Lessens the chance of a major interruption of your daily business.
3. Protects your company's or your brand's reputation.
4. Allows daily business activity to continue with minimal disruption if you are affected.

Companies must establish open lines of communication with all audiences while dealing with the effects of the pandemic or other significant events. Should one occur, these stakeholders will want to know what you are doing to manage the situation and minimize their risks. If you communicate with these stakeholders openly and promptly, you send four valuable messages:

• You are taking charge of the situation.
• You take it seriously.
• You have the best interests of your staff and customers at heart.
• You run a responsible company with nothing to hide.

Pandemic planning begins with human resources

Pandemics have a disastrous effect on a company's optimal functioning because they prevent large numbers of critical employees from showing up for work. The resulting interruption to normal operations can have a disastrous cascading effect, affecting nearly every corner of the organization at considerable cost.

Employees unable to work or prevented from working become anxious and insecure. When they start asking management questions that aren't answered sufficiently or quickly, it exposes the fact that management hasn't developed contingency plans or that management failed to consider what employees need to know. Part of the cost of failing to prepare can be measured by the resultant loss of trust in management's capability, judgment, and credibility.

We know from experience there are certain predictable questions that employees will ask and HR departments must be prepared to answer. For example:

1. Will H1N1 close our business down?
2. If yes, what will happen to my paycheck?
3. How long could we be closed?
4. How long could the company be closed and still survive?
5. What are we doing to make sure we can stay in business?
6. Will I still have a job if the flu forces us to shut down?
7. Will I still get paid if I get the flu and have to stay home?
8. Will I get paid if schools close and I have to stay home with my children?
9. What will happen to my health insurance coverage?
10. What will happen if I run out of sick days?
11. How will I find out what is happening around the company and how it might affect me?
12. If H1N1 hits us, how will my job change? Exactly what will I have to do?
13. Will it be possible for me to work from home, using the Internet and phone?
14. I do not want to be forced to work next to someone who's sick. What is our policy regarding people who insist on coming to work when they have the flu?
15. What should I tell our customers/vendors/partners, etc., when they ask what's going on?

HR departments should, as a matter of urgency, review attendance and sick-day policies to ensure they have made allowances for managing the larger-than-normal issues H1N1 creates. Some of the policies that will need to be considered for implementing or addressing include:

1. How/when to start monitoring/screening employees at the workplace to determine if they are sick or pose a risk.
2. How/when sick employees should be sent home to protect colleagues at work or be stopped/prevented from coming to work where they could infect colleagues.
3. How/when the company should be temporarily closed due to the number of sick employees.
4. How/when to implement steps to minimize face-to-face contact at work.
5. How/when to allow certain employees, including senior management, to work remotely from home or another branch/office.
6. How/when employees should be allowed to stay at home to look after sick family members.
7. How/when the company's travel policies should be changed/suspended.
8. How/when to stop employees from coming into contact with suppliers and customers.
9. How/when to implement and enforce a "wash your hands" and "cover your mouth and nose when coughing and sneezing" policy; this must include making face masks and the use of hand sanitizers mandatory across the company.

10. How/when to change the company payroll policy so that all employees receive electronic payments into their accounts; consider establishing an emergency "employee help" fund.

11. Any and all extensions/additions to your existing payroll and work hours' policies.

At the core of your H1N1 crisis plan, your HR department must be fully prepared to explain and communicate any new policies or changes to employees on an ongoing basis in all offices. This includes offices and employees that may not be affected by the pandemic at all. International and regional offices must also be briefed as they, too, could be directly impacted if there is an H1N1 outbreak.

Employees should also be asked for input and ideas. This may help to highlight potential management or operating aspects that have not been considered. It will also make employees feel part of the pandemic planning process and thus, more accepting of and cooperative with the final plan.

If appropriate to your workplace and organizational culture, additional steps can be taken to protect employees by putting up educational posters, using training materials, and even arranging for annual flu shots (under doctor's supervision) to be provided in the workplace for convenience. Employees should also be encouraged to learn and do more on their own and away from work.

All of these actions send a message to employees that you are looking out for them, their jobs, and the company's well-being. In return, employees are much more likely to "go the extra mile" in order to lessen the business impact of widespread absences.

Communicating during a crisis is important, but what businesses do is always more important than what they say. Making good decisions and providing straightforward, honest and factual information to all employees with frequent updates is one of the most critical actions management can take.

Crisis planning specifically for H1N1

Ideally, all companies and organizations would have enterprise-wide crisis plans in place before a crisis breaks. But realistically, we know from multiple surveys that at least half don't. Too many companies assume an "it can't happen to me" mentality or, in tough business or competitive conditions, they decide not to invest in "insurance" activities. Unfortunately, some find out the hard way that you cannot choose your crisis; it chooses you—and almost always at the most inconvenient time.

If yours is an organization that hasn't taken the steps necessary to implement crisis preparedness, here are some **interim steps that you can take quickly** to address H1N1. Remember, the most effective and least costly way to manage a crisis is to prevent it from happening in the first place. You cannot stop H1N1, but you can take steps to keep it from damaging your operations, your reputation, and your bottom line.

Here's a quick checklist of things an organization can do, even at this late date:

1. **Appoint a pandemic coordinator or team.** This individual or team will lead the organization through various steps to become pandemic-ready.

2. Have them first **conduct a vulnerability and risk assessment.** That means identifying areas in which you are at heightened risk of infection or in which your responses or ability to compensate will probably be weak. Armed with this knowledge, you should be able to prepare for **worst-case scenarios** and begin planning accordingly.

3. Get your **Crisis Management Team** up to speed. A crisis management team consists of senior employees who will deal full time with a crisis while the rest of the organization runs as normally as possible. The most effective crisis teams typically consist of no more than five members who serve as its decision-making leadership. Crises are not situations for committees or consensus building. They demand swift and certain decisions and actions be made under "battlefield conditions." We strongly recommend that you have a "five-star general" heading up your team.

4. A Crisis Management Team must possess sufficient inherent or delegated power to **command unrestricted access to a full cross-section of corporate disciplines**, including HR, sales, customer service, information technology (IT), security, operations, facilities management, communications, department/business unit heads—from every corner of your organization. The Crisis Managers must know who from these disciplines are to be brought on to support the Crisis Management Team on an as-needed "on-demand" basis. Note that these disciplines are for advice and support, not crisis decision making.

5. Management should assign each person on the Crisis Management Team and the designated support providers to **specific roles and functions ahead of time—and** give them full authority to carry them out.

6. The team should also include someone who will be **company spokesperson** throughout the crisis. Ideally, the spokesperson should be a senior company executive. He or she should have received formal media training, and should have the stamina, self-discipline, and inner strength to be able to convey trust and believability when speaking during a time when bad news may need to be delivered to various audiences.

7. Think about including **external experts** on your team. These could include public health consultants, doctors, HR consultants, and business continuity experts.

No organization can hope to be crisis-ready unless it is prepared with **messaging** ready to be disseminated to audiences on short notice and under pressure. Crisis messaging typically consists of fully or partially (fill-in-the-blanks type) prepared statements addressing a range of potential situations **anticipated in advance**. Prepared organizations keep them in a template format. Then, as a crisis develops and the actual facts of the situation become known, the relevant template can be rapidly updated with all pertinent information.

In a crisis, you simply do not have time to agonize for long over "What are we supposed to say?" Remember, it is only during the first 60 minutes of a crisis that you have your one chance to **take control of the situation via proactive communication**. In that time, messages must be disseminated internally to staff and externally to the relevant audiences, such as customers, stockholders, suppliers, and partners, and possibly the media.

Businesses that conduct vulnerability and risk assessments will have a better idea of the templates and draft messaging they will need for a flu outbreak. These situations range from temporarily closing a site to announcing an interruption of service. The tone of all messaging must demonstrate that management is taking the situation seriously.

Employees are your first priority and must receive crisis-related messaging before anyone else. The media and relevant external stakeholders can then receive the same or similar messaging soon after. Department heads in your company can be used to communicate directly with employees. Employees should also be provided with messaging that they can share with others outside the organization. In today's "always-on" instantaneous online world, whatever employees are told invariably becomes public knowledge within minutes.

From time to time, someone will ask a question that cannot be answered using prepared messaging. The crisis team must be prepared to reply "I don't know," and then either explain why, honestly and plainly, or commit to providing the answer at a given time in the future. Nothing destroys trust and creates anger more than speculating or guessing at answers that may be proven wrong at a later stage. While you must *respond* quickly to all questions, you may not be able to *answer* them all. The crisis team must understand the difference.

Stakeholders want reassurance you are doing everything possible to manage the situation and communicating without a hidden agenda. If you intend to keep your business open and running during a significant event, say so. For credibility, communicate the steps that you are taking to ensure it is kept open. If you are asked questions and are uncertain about what will take place, acknowledge this honestly. Make every effort to find the answer quickly and, when you have it, follow up as soon as possible.

Plan to work with third parties. Adopting a go-it-alone attitude in dealing with a pandemic is needlessly dangerous. Organizations are wise to be working with key third-party consultants to make crisis preparedness as robust as possible. Key third parties could include:

- Crisis public relations (PR) consultants
- Doctors and pandemic specialists
- Public health departments
- Emergency medical responders
- HR consultants
- Lawyers
- Local hospitals
- Red Cross
- Security services

Don't overlook your supply chains. Companies providing each other with operations-critical products, goods, or services become inextricably linked. A problem in another company may cascade to yours, affecting your ability to meet contractual obligations. Steps they take to stay in business may be beneficial or disruptive to you. Knowing ahead of time will help you make appropriate arrangements or establish alternatives. Cooperating with customers, partners, suppliers, and local governments helps you become pandemic-resilient.

Expert legal opinion must be obtained on how to address contractual obligations should a full scale pandemic break out. If you're prevented from delivering products or services and thus break legally binding contracts, customers/partners could hold you liable for failing to plan adequately. Such legal action could expand or precipitate a second crisis, when the media reports the legal action and you are forced to deal with a reputational crisis.

Crisis communications tools for H1N1

During a pandemic, organizations must communicate effectively with all internal and external audiences. Being ready to communicate proactively and at a moment's notice requires advance preparations.

Internal communications

In all cases, employees are the most important communications targets during a crisis. Friends and family will contact them along with many of their external business relationships (including the media) to ask "What's really going on?" And we know from experience that poorly briefed employees tend to speculate in the absence of solid information. This could easily precipitate a secondary crisis, forcing you to deal with rumor-mongering by employees and potentially false reporting by the media. Either could cause serious damage.

Thus, you must **designate in advance your primary or "official" internal communication channels**, and let everyone in your organization know what they are. While face-to-face verbal communication is the best medium for internal audiences during a crisis, it may not be possible if H1N1 strikes. Depending on your specific situation, one of the following channels should be considered in order to communicate companywide:

- Teleconference
- Webcast
- E-mail
- Public address system
- SMS (texting)
- Company intranet
- Blast voicemail
- Call-in hotlines

Remember: What is written and given to employees can be passed on to the media and other parties.

External communications

Communication with all external stakeholders must be **timely and accurate, with messages consistent with what is being communicated internally**. Messaging differences should be determined by relevance to the receiver. But be safe: when in doubt, overcommunicate. In a crisis, everyone wants *more* information, not less.

If you had to **communicate with 100% of your customers within 60 minutes**, could you? Do you have up-to-date accurate contact information housed in databases that can support mass messaging such as blast e-mail or recorded voice messages with outbound autodialing? Blast-fax? Cell phone information for texting?

Nobody has time to build these contact databases once a crisis strikes. **Assemble them now**.

The best time to start communicating is when there is no crisis. A proactive information campaign could spearhead the opening of new channels of communication with your various external audiences prior to a crisis.

External communication channels

The following external communication channels can be used proactively or reactively depending on the situation:

- Company website
- Teleconference
- Webcast
- E-mail
- SMS
- Voicemail
- Faxes
- News releases including wire services
- Call centers (inbound and outbound)
- Electronic signage

While social media tools such as Twitter, Facebook, YouTube, and blogs can play a role in crisis communication, at this time we believe they are not the tools best suited to be your primary or "official" communication channel to the outside world. Especially for business organizations, social media are not yet universally accessible.

But more importantly, they are not within your complete control. You must be extremely careful about what you say via social media, as it is very difficult to change anything after it has been sent out. It's the very nature of most crises that the situations and facts change, and change often. Social media messages containing old information can too easily recirculate, causing misunderstandings and conflicts precisely at a time when they can do the most damage.

Business continuity management

A major H1N1 breakout could devastate supply-and-value chains, and possibly close down entire industry sectors. This will prevent companies from providing or delivering much needed services. Customers, partners, suppliers, and employees will feel a significant impact. There will also be financial repercussions.

In short, a business could be forced to close down if it is not ready for all eventualities.

To be truly resilient in a crisis, the organization must have an up-to-date business continuity plan detailing how it will restore its operating functions, either totally or partially, within a certain period of time.

To achieve this, key decision makers must:

- Have an in-depth look at their company to **identify essential functions** needed to keep doors open. Nonessential ones can be temporarily discontinued without impacting day-to-day operations. People with **key skills** that are important to the business during the pandemic must be identified and protected whenever possible. Those with nonessential skills may be told not to report for work during the pandemic.
- Consider contingency plans to **switch operations to other sites**, if possible.
- **Identify alternative suppliers** that you can switch to at a moment's notice. Your primary suppliers of utilities, goods, products and services may suddenly shut down because of poor planning. You should ask current suppliers to disclose what contingency plans they have in place to ensure the provision of uninterrupted service to you. Put backup plans in place to switch to other/competing suppliers and contractors if you're the least bit unsure of their preparedness.
- Determine if their **IT systems are sufficiently robust** so critical technology-dependent business processes would still function.

References

AON. (May 2009). *Exploding the myths: Pandemic influenza* (2nd ed.). Chicago, IL. Available at <http://img.en25.com/Web/AON/H1N1_WP_050409_72dpi.pdf> Accessed 16.09.09.

Center for Disease Control (CDC). (August 19, 2009). Preparing for the flu: A communication toolkit for businesses and employers. Atlanta, GA. Available at <http://www.cdc.gov/H1N1flu/business/toolkit/pdf/Business_Toolkit.pdf> Accessed 17.09.09.

Center for Infectious Disease Research and Policy (CIDRAP) 10-Point framework for pandemic influenza business preparedness 2007 CIDRAP Business Source, University of Minnesota last updated September 8, 2006. Available at <http://www.cidrapsource.com/source/index.html> Accessed 15.09.09.

Department of Homeland Security. (September 2009). Planning for 2009 H1N1 influenza: A preparedness guide for small business. Washington, D.C. Available at <http://www.pandemicflu.gov/professional/business/smallbiz.html> Accessed 17.09.09.

Goulet, D. (September 11, 2009). Pandemic planning and your supply chain. Huddersfield, England. Available at <http://www.continuitycentral.com/feature0699.html> Accessed 16.09.09.

Harvard School of Public Health. (September 9, 2009). Four-Fifths of businesses foresee severe problems maintaining operations if significant H1N1 flu outbreak. Boston, MA. Available at <http://www.hsph.harvard.edu/news/press-releases/2009-releases/businesses-problems-maintaining-operations-significant-h1n1-flu-outbreak.html> Accessed 17.09.09.

Mercer, L., & Kapcio, P. Internal communications for the avian flu: Anticipating effects on lives and livelihoods 2006 Public Relations Tactics New York, New York

U.S. Department of Health and Human Services. (January 2006). Pandemic influenza planning: A guide for individuals and families. Washington, D.C. Available at <http://www.hhs.gov/> Accessed 18.09.09.

Air travel health risks and concerns

Even though more than one billion people travel via commercial aircraft every year, illness as a direct result of air transportation isn't common; however, there are risk exposures associated with air travel that both employers and travelers should be cognizant of in order to mitigate the risks when possible.

Air quality within commercial aircraft

Most modern aircraft are equipped with HEPA (high efficiency particulate air) filters, which, according to the European air filter efficiency classification, can be any filter element that has between 85% and 99.9995% removal efficiency. According to Pall Corporation, for aircraft cabin recirculation systems, the definition has been tightened by the aerospace industry to a standard of 99.99% minimum removal efficiency.[5] Most modern aircraft provide a total change of aircraft cabin air 20 to 30 times per hour, passing through these HEPA filters, which trap dust particles, bacteria, fungi, and viruses. Many airlines have an airflow mix of approximately 50% outside air, and 50% recirculated, filtered air whereby the environmental control systems circulate the air in a compartmentalized fashion by pushing air into the cabin from the ceiling area, and taking it in at the floor level from side to side, versus air movement from the front to back of the aircraft. However, most viral respiratory, infectious diseases, such as influenza and the common cold, are transmitted via droplets that are most commonly transmitted between passengers by sneezing or coughing. These droplets can typically only travel only a few feet this way. However, it is their survival rate once they land on seats, seatbelts, tray tables, and other parts of the passenger cabin that can provide additional exposure, which is why sanitation of your personal seating area when traveling, particularly your hands with an alcohol-based sanitizer before eating, is important. Surgical masks have been shown to reduce the spread of influenza in combination with hand sanitization, particularly when worn and practiced by the infected individual.

Viral outbreaks in recent years of concern to business travelers have included Middle East respiratory syndrome (MERS), severe acute respiratory syndrome (SARS), and Ebola, H1N1 (Swine Flu), among others.

The International Air Transport Association (IATA) has developed an "Emergency Response Plan Template" for air carriers during a public health emergency, which can be found at the following link: http://www.iata.org/whatwedo/safety/health/Documents/airlines-erp-checklist.pdf

Disinsection

Disinsection is the use of chemical insecticides on international flights for insect and disease control. International law allows disinsection and the World Health

[5] Pall Corporation, "Cabin Air Q&A," n.d., http://www.pall.com/main/aerospace-defense-marine/literature-library-details.page?id=46181.

Organization (WHO) and the International Civil Aviation Organization suggest methods for aircraft disinsection, which include spraying the aircraft cabin with an aerosolized insecticide while passengers are on board, or by treating aircraft interior surfaces with a residual insecticide when passengers are not on board. Two countries, Panama and American Samoa, have adopted a third method for spraying aerosolized insecticide without passengers on board.

Immobility–blood clots (deep vein thrombosis)

Not specific to just air travel, blood clots or DVT (deep vein thrombosis) can be a serious and potentially deadly health risk for any traveler with restricted mobility in an aircraft, car, bus, or train. Anyone traveling for more than 4 hours without sufficient movement can be at risk. Many blood clots are not necessarily visible and can go away on their own, but when a part of one breaks off, there is the possibility of it traveling to your lungs, creating a pulmonary embolism, which could be deadly.

In addition to traveler training on prevention of DVT, companies should take this threat into consideration with regards to international class of service policies or reimbursement consideration for upgrades.

According to the U.S. Centers for Disease Control (CDC), the level of DVT risk depends on whether you have any other risks of blood clots in addition to immobility, as well as the length or duration of travel.

The CDC also states that most people who develop blood clots have one or more other risks for them, such as:[6]

- Older age (risk increases after age 40 years)
- Obesity
- Recent surgery or injury (within 3 months)
- Use of estrogen-containing contraceptives (e.g., birth control pills, rings, patches)
- Hormone replacement therapy (medical treatment in which hormones are given to reduce the effects of menopause)
- Pregnancy and the postpartum period (up to 6 weeks after childbirth)
- Previous blood clot or a family history of blood clots
- Active cancer or recent cancer treatment
- Limited mobility (e.g., a leg cast)
- Catheter placed in a large vein
- Varicose veins

Civil unrest (including active shooter situations)

Civil unrest generally takes place when a group of people in a specific location is angry, resulting in protests and violence. Around the world, there are countless incidents of civil unrest that erupt, which can not only cause inconvenience and safety concerns for business travelers, but can also cause mental and emotional stress for

---

[6] U.S. Centers for Disease Control (CDC), "Blood Clots and Travel: What You Need to Know," n.d., http://www.cdc.gov/ncbddd/dvt/travel.html.

which the employer is ultimately responsible to try to limit the effects of whenever possible, and to treat as early as possible after the incident is over.

Within the first 6 months of 2014, the world saw civil unrest and protests in Turkey, Brazil, Ukraine, Thailand, Venezuela, Malaysia, Cambodia, India, Egypt, Hong Kong, Russia, China, and the United States (excluding military acts of war or civil war).

In January of 2011, governments and private organizations from around the world began evacuating people from Egypt due to civil unrest. Approximately 50,000 Americans lived and worked throughout Egypt at the time, and approximately 2400 requested evacuation assistance from the U.S. Government. Such an exercise requires massive planning and resource availability, even for much smaller groups of people. Consider the number of other companies competing for the same resources to evacuate their people, as well as the general public trying to leave. Companies without a plan in place, along with proper strategic crisis response resources, would have been last in line to evacuate their impacted travelers and at greater risk for someone getting hurt or killed.

At one time, civil unrest may have been considered primarily politically motivated, but today, there are many factors that lead to the spark that starts the fires of violence. Things such as overpopulation, lack of food and resources, poverty versus wealth (income inequality), crime, lack of jobs and religious persecution, while sometimes related to political causes, are all reasons for the increased violence we see today. With the advent of mobile technology being increasingly available to the middle and lower classes of the world, it doesn't take much or long time-wise, to incite anger or hatred in others who can assemble quickly, sometimes before one has a chance to react. Throughout the text of this book, readers should see a common theme about the importance of quality risk intelligence. The previous statement about violence breaking out before one can react, is a perfect example of how real, risk intelligence (not simply recycled news) can often predict these events as they are starting to come together and warn people in advance, so that companies and individuals can take steps to mitigate their exposure. In such examples, would employers and travelers want "cheap information" from a provider that primarily scrapes news wires on the Internet, or qualified, vetted security analysts with thousands of sources? If a life depended on it, I'm confident that people would choose vetted intelligence. Another way to understand the value of news versus intelligence is that "intelligence" is in effect "analysis + news + context + advice." Experienced security analysts specializing in specific geographic areas and subject matter produce quality intelligence.

Climate change can also drive civil unrest with sea-level risings, damage to property, water shortages, and increased costs associated with lost productivity or infrastructure collapse. People simply go where the goods and the work are provided. When that is lost for various reasons over a large area, there can be mass migrations that sometimes see the intervention of military units to prevent border crossings and an unanticipated drain on other population's resources.

Property damage and serious violence in Vietnam in May 2014, as a result of anti-Chinese protests, was experienced not only by Chinese businesses, but by other assets owned by companies from additional countries. Some manufacturing experienced an interruption to production, causing between 4 percent and 16 percent decreases in company share prices. These figures and insight are intended to support business cases for companies to invest in not just products and programs to avoid business disruptions caused by civil unrest and other factors, but the time required to simply have plans in place to mitigate the risk.

Harassment by authorities

Imagine being in a foreign country on business and getting pulled over on the road in your rental car by a local police officer. Unaware of any laws that you may have broken, after a quick discussion with the officer, you realize that they are extorting you for a bribe and you simply don't have the cash or the training to respond to the situation properly. Alternatively, a traveler arrives in foreign country via a commercial flight, carrying marketing collateral and merchandise to give away at a conference that they are attending. The local customs authorities misinterpret part of your merchandise, because the conference is being held in a deeply religious country with harsh laws regarding morality. Not only does the traveler fear for their safety, the company doesn't want to cause an international incident, which can be difficult to clean up. Does your company provide resources and training to travelers regarding how to handle themselves in such situations?

Considerations for female travelers

Women from Western countries may still find it hard to believe how many places in the world where their personal safety, and possibly their lives, can depend upon the length of their skirt and sleeves, or the time of day that they are out and about, particularly without a male escort. In 2013, a woman from New York was found dead in Turkey; a Turkish man confessed to killing her after allegedly trying to kiss her. According to news reports, she was a first-time international traveler, an avid social media user, and was in constant contact with friends and family. It is reported that she wasn't off the beaten path or doing anything risky, simply taking photographs.

Sometimes just having some awareness training about your destination can save female travelers the potential for conflict or incident, such as holding one's purse in her lap or at her feet with a thick strap around her leg to secure it, or ensuring that luggage tags do not openly display addresses and have a cover that must be opened to reveal the information. According to Joni Morgan, Director of Analytic Personnel at iJET International, "In some cultures, for instance, it's not appropriate for a woman to initiate a handshake." "In Afghanistan, it's considered an insult to show the bottom of your shoe, so when crossing your legs, you want to be aware of that."[7]

Female road warriors are learning important skills that are notably helpful in all destinations, but in some more than others, additional care should be taken. Indications of when to take additional care is an important part of pretrip travel

[7] Charisse Jones, "Female Business Travelers Face Special Challenges," *USA Today*, March 8, 2015, http://www.usatoday.com/story/travel/roadwarriors/2015/03/08/tips-for-women-business-travelers/23889099/.

intelligence provided by an employer's TRM program, supported by a vetted travel risk intelligence provider.

Some considerations for female business travelers while traveling alone or even with peers on business include the following:

1. Always plan your route before going anywhere. Never leave your hotel or office without understanding where you are going and appropriate routes. Travelers do not want to look lost in the street looking at maps or their mobile devices for directions.
2. Use vetted taxis or ground transportation providers. Make an attempt to prebook all transportation with providers that your company has preapproved, and have appropriate security policies and procedures in place, such as identifiable car numbers, driver identification, tracking, and electronic order confirmation. Removing the potential for unfamiliar, unvetted ground transportation providers can drastically reduce the potential for assault or abduction.
3. Purchase peephole blockers for hotel room doors. For a small amount of money, travelers can purchase a device to block the outside view of the inside of their hotel room by assailants who have devices that enable broad visibility inside hotel rooms from the outside via peepholes. In the absence of such a device, place tape or a sticker over the inside peephole opening.
4. Choose your hotels carefully. Make it clear to your employer that you take safety seriously and that you expect safety considerations to have been taken into account when designating preferred hotels for employees to stay at. Employers should be able to articulate what kinds of safety standards go into their preferred hotel selections, which form the basis for how different incidents can be mitigated or handled should an incident occur.
5. Never stay at hotels or motels where the room door is exposed to the open air (outside).
6. Try to not accept hotel rooms on the ground floor. Being on a higher floor makes it more difficult for an assailant to get away or not be seen on surveillance cameras.
7. Never tell anyone your room number verbally. If a hotel employee asks for it, provide them with it in writing and personally hand it to them. Do not write it on a check and leave it unattended. You don't want someone in the area to overhear you providing this information verbally or to view it on your check.
8. Alcohol consumption—Never leave your drink unattended or out of your sight. A momentary distraction is an opportunity for someone to place drugs into your drink. Also, never drink until intoxicated while on business and be mindful of locations where drinking alcohol may even be illegal.
9. Emergency phone numbers—Know the equivalent of 911 or the local emergency services phone number and your local consulate or embassy phone numbers and preprogram them into your mobile phone, in addition to your company's provided crisis response hotline. Whichever number you are instructed to call first according to your company's policies (if your company provides a crisis hotline), having those numbers handy can save your life when moments count.
10. Never tell anyone that you are traveling alone. Avoid solitary situations. Try to remain in social situations where plenty of people are around. If you feel uncomfortable, leave.
11. Leave a TV or radio on when you leave your hotel room to provide the perception that someone is in the room.
12. Never hesitate to ask security or someone to escort you to your room, and avoid exiting an elevator on your hotel room's floor when sharing the elevator with a man. If necessary, go back to the lobby level until more people get on the elevator or you can ride it alone.

13. Use valet parking. Self-parking can often put individuals at risk of assault in unsupervised car parks or garages.
14. Upon arrival at your hotel, take a hotel business card or postcard and keep it with you at all times. If ever you are away and need to return, and you either don't remember the address, or your driver doesn't know where it is, or you don't have a signal on your mobile device, you can use the card to provide address details (usually in the local language).
15. Do not use door-hanging room service order forms (typically for breakfast), as they often note how many guests you are ordering for.
16. Make sure you have adequate insurance. Just because you are on a business trip, doesn't mean that your employer has obtained enough insurance or services to support you in the event that a crisis occurs. Hopefully, employer-provided insurance and support services are adequate and have been effectively communicated, but don't travel for business without a thorough understanding of what kind of coverage and support you have. In particular, any medical coverage should guarantee advance payment to local service providers and not require travelers to pay for services and file for reimbursement upon their return home. Most people don't have access to the many thousands of dollars that might be necessary to procure sufficient treatment and support.
17. Travel with smart travel accessories. Travel with a small, high-powered flashlight and one or more rubber door stops for the inside of your hotel room (be aware of the downside of using in case of a fire).
18. Leave copies of your passport with someone at home who can easily get a copy to you if you need it. Having a copy can expedite the replacement of a lost or stolen passport if needed.

Cultural or social stigmas and violence against women

Honor killings

An honor killing is a homicide of a family member, typically by another family member, based upon the premise that the victim has brought dishonor or shame to the family, in such a way that violates religious and/or cultural beliefs. Again, as with religious or cultural restrictions on modest clothing, honor killings are not exclusive to women, but within the cultures and countries where honor killings are more generally accepted, men are more commonly the sources or perpetrators of the revenge or honor killings, very often charged by the family to watch over and police female family member behavior, restricting or prohibiting things such as adultery, refusal to accept an arranged marriage, drinking alcohol, or homosexuality. Honor killings are not exclusive to any one country or religious faith, because they are found in a broad scope of cultures, religions, and countries. Although more common in places such as the Middle East and Asia, there have been documented cases of honor killings in the United States and Europe.

If honor killings were based largely on the premise of family honor, why would nonfamily members or business travelers need to be concerned? Honor killings have been known to happen to nonfamily members in strict, culturally conservative countries. Perceived inappropriate behavior, typically with a female member of a conservative family, could result in the killing of the female family member and the nonfamily suspect. Such killings can even take place in broad daylight. In Lahore, Pakistan in 2014, one such incident occurred involving multiple participants while the police looked on. The victim killed for marrying a man that she loved without family consent.[8] Often these crimes are hard to document or record because they are disguised as suicides or, in some Latin American countries, as "crimes of passion." The United Nations Fund for Population Activities (UNFPA) estimates that as many as 5000 women fall victim to honor killings each year.[9]

Dress expectations for women

Article 57 of Qatar's constitution states that it is a "duty of all" who resides in or enters the country to "abide by public order and morality, observe national traditions and established customs." This means that wearing clothing considered indecent or engaging in public behavior that is considered obscene is prohibited to all, including visitors. In Qatar, the punishment could be a fine and up to 6 months in prison. With kissing or any kind of physical intimacy in public, as well as homosexuality, being outlawed under Sharia law, *all* travelers to or via the Middle East for business or tourism purposes (e.g., to attend the 2022 World Cup), should take heed.

The Qatar Islamic Cultural Centre has launched the "Reflect Your Respect" social media campaign to promote and preserve Qatar's culture and values. Posters and leaflets advise visitors, "If you are in Qatar, you are one of us. Help preserve Qatar's culture and values, please dress modestly in public places." While research finds no definition in Qatar's Article 57 for modest clothing, campaigns such as this suggest that people cover up from their shoulders to their knees and avoid wearing leggings. They are not considered pants or modest dress. An example of the campaign leaflet can be found in "Qatar Launches Campaign for 'Modest' Dress Code for Tourists" published by the *Independent* (UK newspaper).[10] Modest dress applies to both men and women. Of course, strict laws, preferences or rules regarding dress expectations for women are not exclusive to any one country. http://www.pewresearch.org/fact-tank/2014/01/08/what-is-appropriate-attire-for-women-in-muslim-countries/.

Sexual assault, harassment, and objectification

While each employer may have specific approaches to handling an incident such as sexual assault, there must be a defined process for reporting such an event that involves crisis response resources that can intervene and provide advice on how to handle the situation with local authorities, perhaps first by contacting diplomatic contacts before contacting the police. Facing local authorities alone in a foreign country for such a sensitive issue as sexual assault can be daunting and intimidating

---

[8] NBC News, "Family Stones Pakistani Woman to Death in 'Honor Killing' Outside Court," May 27, 2014, http://www.nbcnews.com/news/world/family-stones-pakistani-woman-death-honor-killing-outside-court-n115336.

[9] United Nations, Resources for Speakers on Global Issues, "Violence Against Women and Girls: Ending Violence Against Women and Girls," http://www.un.org/en/globalissues/briefingpapers/endviol/.

[10] Lizzie Dearden, "Qatar Launches Campaign for 'Modest' Dress Code for Tourists," *Independent*, May 27, 2014, http://www.independent.co.uk/news/world/middle-east/qatar-launches-campaign-for-modest-dress-code-for-tourists-9438452.html.

without a company or diplomatic representative being there to assist. Crisis response suppliers should be equipped with necessary contacts, recommended protocols, and resources to help the victim and employer to address the situation and get help as soon as possible. This is another good example of why employers should have a single global crisis response hotline for any crisis that a traveler may encounter while on business travel.

Sexual harassment can happen anywhere. What happens if you require a traveler to use a supplier per the company's travel policy, and a representative of that supplier sexually harasses the traveler? In addition to standard protocols within the workplace, considerations must be given to business travel, which from many perspectives today is an extension of the workplace.

Sexual Harassment Case Study

A female business traveler, over the course of several months on a project, travels during the week, returning home on weekends. Over time, a car rental clerk at the location she rented from weekly, began making comments to her about her appearance each time she checked-in or returned a car. Eventually, the rental clerk began calling her mobile phone to share how he liked what she was wearing and began sending her text messages while she was in town, using the mobile number she provided at check-in. Not responding and scared, the traveler canceled all future reservations and books rental cars with another provider. Shortly thereafter, the clerk began calling and texting her, asking why she canceled and when she would be coming back.

A concerned colleague of the traveler brought the situation to the company's travel manager, who intervened with their human resources and legal departments to proactively address the situation with the authorities and the supplier, and to provide appropriate support for the traveler as best they could. The end result, after much investigation, was the issuance of restraining orders against the clerk and termination of his employment. It turned out that the supplier hadn't done sufficient background checks on its employees and the clerk in question had a history of similar behavior.

Hate crimes

A hate crime is a criminal act of violence targeting people or property that is motivated by hatred or prejudice toward victims, typically as part of a group, based upon creed, race, gender, or sexual orientation.

A critical component of any TRM program is disclosure of potential risks to the traveler prior to taking a trip to a destination. In consideration of laws and cultural beliefs in select countries or regions that sanction the persecution, imprisonment or killing of members of the LGBT (Lesbian, Gay, Bisexual, and Transgender) community, specific races, religions, or sex (mainly women), travelers must be prepared with information and training on acceptable behavior when traveling to these destinations and understand how to get help should they find themselves in a difficult position or a potential victim of a hate crime. Saying the wrong thing, at the wrong time, in the wrong place, or wearing something inappropriate, or acting a certain way that isn't culturally acceptable in some parts of the world, can put travelers in real danger. How does your company prepare your travelers for facing these challenges as they travel?

While some laws that promote discrimination that can lead to hate crimes are more notable in the press, such as the antigay propaganda law put into place in Russia prior to the Sochi Olympics, some are less obvious to the average business traveler, such as up to 14 years in prison in Nigeria for simply being gay, or India's Supreme Court ban on gay sex, or the execution of homosexuals in Saudi Arabia.

In April 2013, an 82-year-old man wearing Islamic dress was attacked and killed while walking home from his mosque in Birmingham, UK, by a 25-year-old Ukrainian student who told police that he murdered the victim because he hated "nonwhites."[11]

According to "One in Six Gay or Bisexual People Has Suffered Hate Crimes, Poll Reveals," a 2013 article in the *The Guardian* (UK), some 630,000 gay and bisexual people in the UK have been victims of hate crimes in the previous 3 years, prompting police to take the problem more seriously.[12]

Such examples continue to support the notion that a crisis doesn't need to be an incident that impacts large numbers of people at once. Quite often they involve one person at a time, and they don't need to take place in a high-risk destination, thus discounting the argument by some companies that TRM isn't necessary for those who don't travel to high-risk destinations. A crisis can happen anywhere for many different reasons, affecting as few as one person at a time.

Anti-LGBT laws and cultural acceptance of violence

Although privacy laws generally prohibit companies from asking employees about sexual orientation, making sure that all employees (of any sexual orientation) understand the dangers that face LGBT travelers, can help to mitigate risks for themselves (if LGBT, traveling with an LGBT person, or if perceived as LGBT) or their fellow travelers, considering that there are many countries still in the world where homosexuality is a crime.

---

[11] Ben Flanagan, "Saudi Woman Killing Sparks Hate Crime Fears in UK," Al Arabiya News, June 19, 2014, http://english.alarabiya.net/en/perspective/features/2014/06/20/Saudi-woman-killing-sparks-hate-crime-fears-in-UK-.html.
[12] Jamie Doward, "One in Six Gay or Bisexual People Has Suffered Hate Crimes, Poll Reveals," *The Guardian*, October 12, 2013, http://www.theguardian.com/world/2013/oct/13/one-in-six-gay-people-hate-crimes

Amnesty International Facts and Figures

Making love a crime: criminalization of same-sex conduct in sub-Saharan Africa

- Homosexuality is still illegal in 38 African countries (Algeria, Angola, Benin, Botswana, Burundi, Cameroon, Comoros, Egypt, Eritrea, Ethiopia, Gambia, Ghana, Guinea, Kenya, Lesotho, Liberia, Libya, Malawi, Mauritania, Mauritius, Morocco, Mozambique, Namibia, Nigeria, Sao Tome and Principe, Senegal, Seychelles, Sierra Leone, Somalia, South Sudan, Sudan, Swaziland, Tanzania, Togo, Tunisia, Uganda, Zambia, Zimbabwe).
- There is no criminal law against homosexuality in 16 African countries (Burkina Faso, Cape Verde, Central African Republic, Chad, Congo-Brazzaville, Cote d'Ivoire, Democratic Republic of Congo, Djibouti, Equatorial Guinea, Gabon, Guinea-Bissau, Madagascar, Mali, Niger, Rwanda, South Africa).
- In Mauritania, Sudan, northern Nigeria, and southern Somalia, individuals found guilty of "homosexuality" face the death penalty.
- The last five years have witnessed attempts to further criminalize homosexuality in Uganda, South Sudan, Burundi, Liberia, and Nigeria.
- Cape Verde decriminalized homosexuality in 2004, and since 2009, Mauritius, Sao Tome and Principe, and the Seychelles have also committed to decriminalizing homosexuality.
- South Africa has seen a number of positive legal developments over the past decade, including allowing joint adoption by same-sex couples in 2002, introducing a law on legal gender recognition in 2004, and equal marriage for same-sex couples in 2006.
- South Africa has also seen at least seven people murdered between June and November 2012 in what appears to be targeted violence related to their sexual orientation or gender identity. Five of them lesbian women and the other two were non gender-conforming gay men.
- In Cameroon, Jean-Claude Roger Mbede was sentenced to three years in prison for 'homosexuality' on the basis of a text message he sent to a male acquaintance.
- In Cameroon, people arrested on suspicion of being gay can be subjected to forced anal exams in an attempt to obtain 'proof' of same-sex sexual conduct.
- In most countries, laws criminalizing same-sex conduct are a legacy of colonialism, but this has not stopped some national leaders from framing homosexuality as alien to African culture.
- A cave painting in Zimbabwe depicting male–male sex is over 2000 years old.
- Historically, woman–woman marriages have been documented in more than 40 ethnic groups in Africa, including in Nigeria, Kenya, and South Sudan.
- In some African countries, conservative leaders openly and falsely accuse LGBTI (lesbian, gay, bisexual, transgender, and intersex) individuals of spreading human immunodeficiency virus (HIV)/acquired immune deficiency syndrome (AIDS) and of "converting" children to homosexuality and thus increasing levels of hatred and hostility towards LGBTI people within the broader population.
- LGBTI individuals are more likely to experience discrimination when accessing health services. This makes them less likely to seek medical care when needed, making it harder to undertake HIV prevention work for, and to deliver treatment where it is available. In many government programs they are not identified as an "at risk"

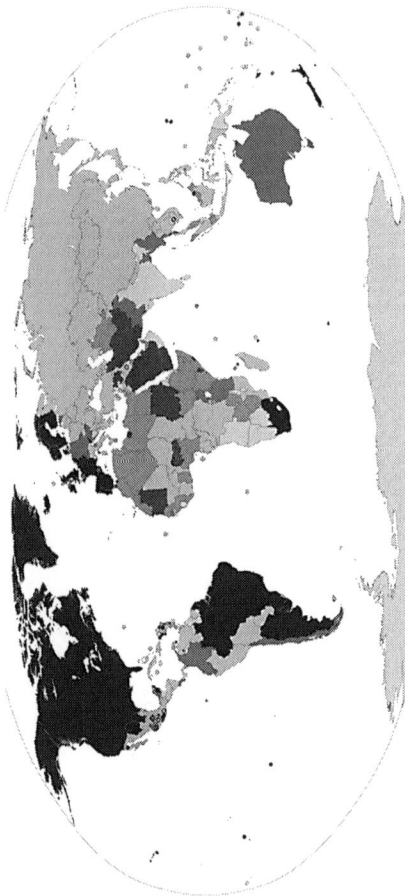

Homosexuality legal

- Same-sex marriage[1]
- Other type of partnership (or unregistered cohabitation)[2]
- Foreign same-sex marriages recognized
- Limited recognition of same-sex marriages at the federal level, no state level recognition
- No recognition of same-sex couples

Homosexuality illegal/restrictions

- Laws restricting freedom of expression and association[3]
- De jure penalty that is de facto not enforced
- Imprisonment
- Imprisonment (up to life sentence)
- Up to death

Rings indicate areas where local judges have granted/denied marriages or imposed the death penalty in a jurisdiction where that is not otherwise the law and/or areas with a case-by-case application.

[1]Some jurisdictions in this category may currently have other types of partnerships or recognize foreign same-sex marriages.

[2]Registered unions in Estonia go into effect on 1 January 2016.

[3]Other countries with similar laws are shown in darker colours if they also criminalized same-sex relationships.

Map regarding restrictions or legality of homosexuality. Countries with laws criminalizing homosexuality are Afghanistan, Algeria, Angola, Antigua and Barbuda, Bangladesh, Barbados, Belize, Bhutan, Botswana, Brunei, Burundi, Cameroon, Comoros, Dominica, Egypt, Eritrea, Ethiopia, Gambia, Ghana, Grenada, Guinea, Guyana, India, Iran, Jamaica, Kenya, Kiribati, Kuwait, Liberia, Libya, Malawi, Malaysia, Maldives, Mauritania, Morocco, Myanmar, Namibia, Nauru, Nigeria, Oman, Pakistan, Papua New Guinea, Qatar, Saint Kitts and Nevis, Saint Lucia, Saint Vincent and the Grenadines, Samoa, Saudi Arabia, Senegal, Seychelles, Sierra Leone, Singapore, Solomon Islands, Somalia, South Sudan, Sri Lanka, Sudan, Swaziland, Syria, Tanzania, Togo, Tonga, Trinidad and Tobago, Tunisia, Turkmenistan, Tuvalu, Uganda, United Arab Emirates, Uzbekistan, Yemen, Zambia and Zimbabwe.

Source: Eva Cantarella, Bisexuality in the Ancient World (Yale University Press, 1992, 2002, originally published 1988 in Italian), p. xi; Marilyn B. Skinner, introduction to Roman Sexualities (Princeton University Press, 1997), p. 11, as cited in Wikipedia, "LGBT Rights by Country or Territory," note 11, https://en.wikipedia.org/wiki/LGBT_rights_by_country_or_territory#cite_note-11

group, and therefore are not catered for in national HIV treatment and prevention programs. As a result, many are denied access to crucial treatment for HIV and other medical issues.

- In South Africa's Gauteng province, 7.6 percent of black gay men and 8.4 percent of black lesbians reported being refused medical treatment because of their sexual orientation. Men who have sex with other men are nine times more likely to contract HIV than other men. Additionally, an LGBTI activist arrested in Uganda in 2008 was denied medical care for diabetes while in custody.
- Arrests for same-sex conduct have been on the rise in the past decade as more regressive policies are enacted. In Cameroon, where there have been 51 documented arrests for same-sex conduct since 2005, people are often detained for up to 48 hours and forced to submit to anal examinations. In Uganda, the harsh 2009 Anti-Homosexuality Bill has led to more cases of people turning their friends and neighbors in to the authorities than previously.
- Lesbians are more often deliberately targeted for sexual violence. Some deem this practice "curative" or "corrective" rape, laboring under the belief that if the victim has sex with a man, she will be "cured" of being a lesbian. Lesbian girls and women in Cameroon can be forced into heterosexual relationships and condemned to double lives. A member of the Cameroonian national soccer team was kicked out of school under lesbian suspicions. Seven lesbians were arrested at a September 2009 Soweto, South Africa pride event and abused in police custody.

Source: http://www.amnestyusa.org/sites/default/files/making_love_a_crime_-_facts__figures.pdf

Kidnapping and ransom

Kidnapping and ransom activities targeting military enemies and employees of multinational companies who are from countries considered to be enemies to terrorist causes, are the primary fundraising strategies of organized terrorist groups. Even for companies that do not routinely visit high-risk locations, having some sort of policy in place for proof of life, which is the means for verifying that a captive is in fact who the captors say they are and that the captive is still alive, such as by providing information that only the alleged victim would know, can save valuable time in a sensitive situation and perhaps someone's life. Additionally, a kidnap and ransom insurance policy is something for all companies to consider, with an understanding that kidnappings happen at anytime around the world, and largely go unreported. According to *The Guardian News and Media* (UK), approximately 75% of Fortune 500 companies have kidnap and ransom (K&R) insurance. K&R insurance originates from 1932, when it was first offered by insurance provider Lloyd's of London, after the kidnapping and murder of American aviator Charles Lindbergh's infant son.

In 2015, the UK's Home Secretary, Theresa May, supported and passed the UK's "Counter-Terrorism and Security Act of 2015," which prohibits insurers from paying claims used to finance payments to terrorist groups. The UK is where many of the world's K&R insurers operate. Many insurers insist that it shouldn't matter because they claim to not pay or finance ransoms, but instead pay claims for services and expenses related to negotiating the release of the captives in question, medical and counseling treatment, along with things such as employee salaries while in captivity. It's difficult to obtain information from clients who hold such policies, because most policies have strict cancelation provisions to prevent a company from disclosing the fact that it has such a policy. Details specific to restrictions on insurance related payments associated with terrorist related ransoms in the UK's Counter-Terrorism Act of 2015 can be found at http://www.legislation.gov.uk/ukpga/2015/6/section/42/enacted.

Companies with any travel to high-risk destinations have a responsibility to provide some kind of survival training for those travelers, in addition to access to resources and provision of current intelligence before, during, and sometimes after their travel is complete.

To complicate matters, based upon a 2013 G8 summit, an agreement was made to not pay ransoms to kidnappers for fear that the money was directly funding terrorist organizations; therefore, some countries, such as the UK, are enacting laws to prohibit the transfer of funds for hostages in certain circumstances or locations. Senior Foreign and Commonwealth Office (FCO) officials in the UK estimate over $60 million has been paid in ransoms to terrorists during the 5 years leading up to the 2013 report. It isn't safe to assume that your government will help bankroll your hostages' release if you find yourself in such a situation, and you may face criminal prosecution if you offer a ransom to specific groups.

Who is at risk?

People who commit kidnappings do so for a variety of reasons, including political or religious views, but most often they are purely financially motivated. Perception is everything, so identifying traveling employees of large or multinational companies, makes them an easy target, thus the reason for using code names for arriving ground transportation signs. Of course, how one dresses and where one goes, also have an impact on how victims are targeted (i.e., wearing expensive jewelry, standing out from the crowd in expensive clothing or making it clear that you work for a large multinational company [clothing with logos or meeting drivers with company names on greeting placards]). Later in this book, kidnappings are explored in greater detail. Some statistics will be presented that both companies and travelers should find serious enough to change their perception about the possibility of kidnapping happening to them. Kidnapping incidents should be accounted for in *all* corporate crisis response plans.

Medical emergencies, evacuations, and insurance

While some medical emergencies may require the need for evacuation, it is more common to receive calls for assistance involving acute or preexisting conditions that can be diagnosed and treated locally. Lost or stolen medication, allergic reactions to food or the environment, and unexpected illnesses, are common occurrences when calling a corporate crisis response hotline. However, in some instances, individuals must be quickly assessed to determine if adequate medical care can be obtained locally, and if not, a decision must be made to evacuate that person to the closest logical facility capable of treating the individual.

Many domestic health insurance plans do not provide coverage for individuals traveling abroad, and often when they do, they require out of pocket expenditures for services; in other words upfront payment by the patient, leaving the patient to file for reimbursement upon the patient's return. More often than not, in these circumstances, this equates to thousands of dollars that most people do not have immediate access to, especially on short notice.

The CDC recommends that if domestic U.S. coverage applies, and supplemental coverage is being considered, the following characteristics should be considered when examining coverage for planned trips:

- Exclusions for treating exacerbations of preexisting medical conditions.
- The company's policy for "out of network" services.
- Coverage for complications of pregnancy (or for a neonate, especially if the newborn requires intensive care).
- Exclusions for high-risk activities such as skydiving, scuba diving, and mountain climbing.
- Exclusions regarding psychiatric emergencies or injuries related to terrorist attacks or acts of war.
- Whether preauthorization is needed for treatment, hospital admission, or other services.
- Whether a second opinion is required before obtaining emergency treatment.
- Whether there is a 24-hour physician-backed support center.

Additionally, one should have coverage for repatriation of mortal remains, should someone covered unfortunately die while away from their home country.

Because so many domestic healthcare plans do not provide for international coverage and evacuations services, companies must provide comprehensive coverage for their employees globally and employees should be fully aware of what is included in said coverage. Employees may decide that what the company offers is not enough by their personal standards and consider purchasing additional coverage to supplement what the company provides. When purchasing different types of travel-related insurance, it's important to understand the differences between the different products offered in the marketplace, especially the differences between consumer and business travel products. Options can include:

1. Travel insurance, which provides trip cancellation coverage for the cost of the trip, delays or interruptions, and lost luggage coverage. It can and often does provide some amount of emergency medical and evacuation coverage, but often requires payment of medical expenses by the insured in the country where services are rendered (versus direct payment by the insurer), and the filing of paperwork for reimbursement upon the insured's return home. Buyers should be mindful of whether or not the policy provides guaranteed payment directly to the suppliers in question.
2. Generally, some consumer based travel health insurance pays for specified or covered emergency medical expenses while abroad; however, such insurance (and others) may require that the individual pay any medical expenses in the country where services are rendered and file for reimbursement upon the individual's return home. Insured parties should always check whether guaranteed payment to providers is included in coverage, as with some consumer-based travel insurance.

3. Medical evacuation coverage is for medical transport to either the closest available treatment facility or the insured's home country for medical attention, depending upon the policy and the situation or medical condition. Considering the cost of medical evacuations, depending upon the distance and the services required for the transport, expenses can vary greatly, but can be very costly. It is recommended that policies have greater than US$100,000 in coverage (some provide up to US$500,000 or more), and include transportation support for an accompanying loved one or family member. Policies with less than US$100,000 in coverage should be reconsidered for possibly not providing enough coverage. Buyers should note that these products cover primarily just the evacuation and not medical services or treatments.

4. Medical membership programs can cater to individual travelers on a per-trip or annual basis or on a companywide basis. These programs can vary widely by provider and membership type, but can potentially provide access to network services resources with separate liability for payment, or network access with some coverage for payment of specified services rendered based upon premiums and policy guidelines.

United States–Workers' compensation

The LII at Cornell University Law School provides a third-party overview of workers' compensation.[13] Variable forms of this type of coverage are provided at both the state and federal levels in the United States, with similar forms of workers' compensation laws also in place in select countries around the world. These laws are typically intended to provide some form of medical benefits and wage replacement for employees who are injured on the job. This coverage is often provided to employees in exchange for releasing their right to sue their employer for negligence, sometimes with fixed limits on payment of damages. Employers need to understand whether the workers' compensation coverage that is applicable and in place for their and their employees' protection, covers international travel. In some cases, additional policies or riders will be required to provide coverage for travel outside of the traveler's home country or state. Additional considerations to this kind of coverage should be as to "when" and "where" the coverage is in effect outside of a company office or facility (e.g., business travel). In some cases this may limit employer liability, but whether it does varies by jurisdiction and circumstance. Considering how workers' compensation benefits have been reduced in recent years, especially in the United States,[14] much consideration needs to be given to assessing what coverage is needed for traveling employees above and beyond workers' compensation, and coordinated with crisis response protocols and risk management support providers for efficient case management, claims, and documentation.

All of these considerations provide a strong business case for why employers should have unique and specific programs in place for medical services and evacuations for employees and contractors traveling abroad *in addition* to their standard

---

[13] LII, Cornell University Law School, Wex, "Workers Compensation," https://www.law.cornell.edu/wex/workers_compensation.
[14] Michael Grabell and Berkes, Howard, "The Demolition of Workers' Comp," *ProPublica*, March 4, 2015, http://www.propublica.org/article/the-demolition-of-workers-compensation.

domestic health care plans and workers' compensation plans. No traveler should embark on a business trip without the complete confidence that medical coverage and resources not requiring their personal, out-of-pocket expenditure is being provided by their employer.

Natural disasters

A 2014 study that included disclosures from 767 institutional investors, representing US$92 trillion in assets, provided by sustainable-economy nonprofit gross domestic product (GDP), stated that in addition to increased physical risks that are being caused by climate change, climate change is already impacting their bottom line. One major UK retailer has stated that 95 percent of its global fresh produce is already at risk from global warming. According to the French Foreign Minister, commenting at a 2015 UN conference in Japan, two-thirds of disasters stem from climate change. Comments were made days after the 4-year anniversary of the Fukushima nuclear disaster that killed approximately 19,000 people in 2011 from an earthquake and tsunami. Margareta Wahlstrom, the head of the UN Disaster Risk Reduction Agency, stated that preventative measures provided a very good return as compared to reconstruction. UN Secretary General Ban Ki-moon asked world nations to spend US$6 billion dollars a year on prevention. An important aspect of both a company's TRM and business continuity plan is to determine what are the unique dangers or risks associated with where your offices or facilities are located, as well as where you travel to on a regular basis, making emergency evacuation and safety plans in the event that a unique incident occurs, such as the following case study related to the 2011 Japanese Earthquake and Tsunami. It is important to know what local governments have made available in close proximity to your travelers' or expats' locations in terms of resources, or something that your company itself may provide, such as "vertical evacuation points" to escape rising tsunami flood waters. These vertical evacuation points may be in a building that is tall enough to support large numbers of the local population at a high water level, with ample support systems and supplies. Not understanding and communicating these plans to your people when appropriate could exact a cost in lives, money, and corporate reputation.

Japanese Earthquake and Tsunami

On March 11, 2011, a 9.0 magnitude earthquake created a 124-foot tsunami. More than 19,000 people died or were presumed dead, with more than 400,000 people evacuated and more than 12.5 million people impacted across the country.*

*American Red Cross, "Japan Earthquake and Tsunami: One Year Update, March 2012," http://www.redcross.org/images/MEDIA_CustomProductCatalog/m6340390_JapanEarthquakeTsunami_OneYear.pdf.

Iceland Eyjafjallajökull Eruption and Ashcloud

For the first time in more than 190 years, Iceland's Eyjafjallajökull Volcano erupted on March 20, 2010, with massive lava flows and ash clouds that closed most of Europe's commercial air space for several days, but then the ashcloud spread to other parts of the world, stranding millions of air travel passengers. Based upon the composite map from the London Volcanic Ash Advisory Centre for the period April 14 to 25, 2010, one can clearly see the massive geographic scale of this incident, and why almost all commercial and private air transportation was prohibited and severe shortages of lodging and emergency shelters occurred.

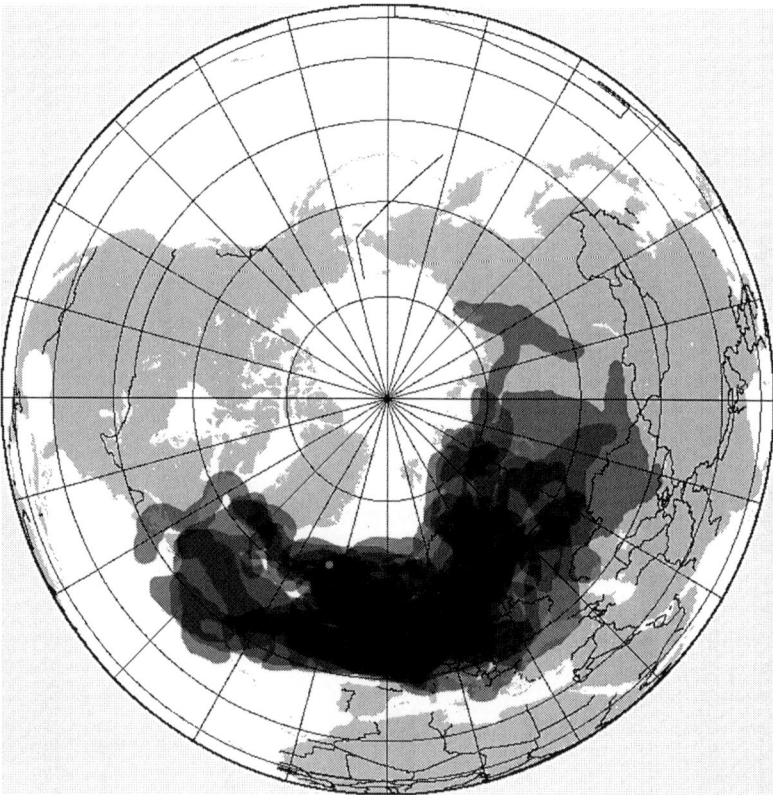

Composite of ash cloud coverage from Iceland's Eyjafjallajökull Volcano eruption during the period April 14–25, 2010.
Source: https://en.wikipedia.org/wiki/2010_eruptions_of_Eyjafjallajökull#/media/File:Eyjafjallajökull_volcanic_ash_composite.png

Whether or not you believe in climate change and the reasons behind it, the statistics demonstrating the depletion of the world's ice sheets and glaciers, warmer ocean waters, and consistent year-over-year sea-level increases, will touch most multinational companies profoundly in the 21st century. *The New York Times* states that sea levels worldwide are expected to rise 2 to 3 feet by the year 2100, but rates are not occurring evenly worldwide. The *Times'* referenced study states that the Atlantic seaboard could rise by up to 6 feet, with Boston, New York, and Norfolk, Virginia, named as the three most vulnerable areas.[15] If current warming trends and rising sea levels continue, cities such as London, Bangkok, New York, Shanghai and Mumbai could eventually end up under water according to Greenpeace, displacing millions of people and causing massive economic damage.[16]

Consider a weather event the size of 2012's Hurricane Sandy, which tips the scales of expected water levels in a low-lying urban city, and results in the displacement of thousands or millions of people, with your travelers or expatriates stuck in the middle of it. When evacuation is not an immediate option, questions regarding the availability of safe accommodation, power, food, and water become priorities as demand far outweighs supply under such circumstances. These occurrences are much more common now than in our recent past.

Evacuations for the disabled

Whether working in their local office or manufacturing facility, or traveling for business, many companies have employees with disabilities. Although building or facility laws and rules may require designated escape routes, ramps, and elevators/lifts in the event of an emergency such as a fire, what about plans for when a disabled traveler is in transit or at a hotel? Special considerations need to be made for disabled travelers in the event of a medical or security-related evacuation, such as:

1. Transporting all medications should the disabled traveler require evacuation.
2. Having adequate medical supplies available during and after evacuation transportation.
3. An accessible method of handicap transport.
4. Addressing any additional criteria needed to determine whether the disabled traveler should be transported or be sheltered in place.
   a. Deciding who makes the call about whether it is safer to "stand by for assistance."
5. Determining whether the transport destination is handicap accessible.
6. Determining whether the transport destination has adequate food, shelter, and supplies for any special needs.
7. Determining whether employers prepared to incur any additional costs relative to evacuating disabled travelers.
   a. Determining whether adequate resources are available.
   b. Identifying the risks or costs for lack of planning.

---

[15] Coral Davenport, "Rising Seas," *The New York Times*, March 27, 2014, http://www.nytimes.com/interactive/2014/03/27/world/climate-rising-seas.html?_r=0.
[16] Greenpeace, "Sea Level Rise," July 4, 2012, http://www.greenpeace.org/international/en/campaigns/climate-change/impacts/sea_level_rise/.

Nonmedical evacuations

The need to relocate travelers can be caused by any number of factors, but before the decision to evacuate is made (usually at considerably more expense than traditional commercial air travel), someone with access to quality intelligence has to make the call as to whether to "shelter in place," assuming safe shelter is available, or to evacuate to the closest safe location. Nonmedical causes for evacuation could be biohazards (e.g., the Fukushima nuclear facility damage in Japan), or civil unrest, or incoming natural disasters. To evacuate or not to evacuate requires thoughtful planning and resources, in order to insure that companies aren't competing with the rest of the world in a reactive situation where many others were caught off guard as well.

---

iJET Case Study—iJET and the South Sudan evacuations

In December 2013, iJET International provided continuous monitoring, intelligence, and analysis of the situation involving heavy ethnic fighting in South Sudan to existing clients with operations in the country. Support included providing real-time situational updates, establishing direct lines of communications with client personnel, and arranging for safe havens and security evacuations. On December 18, 2013, the situation worsened to include the closure of the Juba International Airport. During the first 2 days of fighting, prior to the airport closure, more than 500 people were killed and more than 800 wounded in the violence. During this time, several client personnel traveled across the country's borders to safe havens, but soon after the airport closure, with mounting concerns about large numbers of refugees, those borders quickly closed. iJET successfully evacuated its clients within the first 3 hours of the airport's reopening, bringing in a 15-seat light-passenger aircraft from Nairobi, Kenya, performing some of the first successful group evacuations from this incident without injury.

The iJET case study excerpt is an example of why a company's TRM program cannot consist of technology alone, and discounted news being marketed as intelligence. In situations like these, quality intelligence is what saves people's lives. In this instance, quality intelligence was critical to the coordination of iJET's incident management team's on-the-ground services and support, which lead to not only evacuating its clients, but knowing when was the right time to move its clients to the airport and into the air.

Some medical evacuation services do not provide security-based evacuations, while some can offer both. Companies should consider that one provider for both medical and security services and support, intelligence and insurance, might not always be the best solution. Some companies select one provider for their terms and coverage for medical services, support, and evacuations, but another provider for security-related intelligence, services, and evacuations. There are even those companies with multiple providers for each medical and security service in different parts of the world, working with completely separate insurance providers to pay for the services rendered. Each company must consider the coverage and resources currently available to them via their existing insurance relationship, and then solicit proposals for coverage based upon a clear outline of what the company needs are based upon claims history. Ultimately, companies need a program that can coordinate with all contracted services and insurers, providing a seamless experience for travelers and administrators, and consolidated documentation.

Open bookings

The term "open booking" refers to a booking made by a traveler that was made outside of their managed corporate travel program, avoiding usage of any contracted travel management company (TMC). Technical advances have found ways to incorporate reservations data from multiple websites or suppliers for a traveler's trips into one place for reporting and calendar population. However, to properly capture this data, there are two primary methods available. The first is to allow the applications the ability to scan our inbox for travel-related e-mails and import the data accordingly. The second method is having travelers or independent suppliers e-mail reservation confirmations to an application or "parser," which can parse the data into a standardized database. With some major travel suppliers (such as airlines, for example) there are "direct connections" from their websites to some of these applications. However, in the absence of a direct connection, if you cannot get beyond the security concerns of a third-party application scanning your inbox, one cannot guarantee the automatic capture of 100 percent of open booking data because of human error. For that reason and many others relative to policy and program management, and because of the high probability of human error, for effective TRM, open booking should not be promoted as a primary booking method within a managed travel program. However, there is a place for open booking technology within a managed travel program: to help capture data from travel data normally considered "leakage," which is often not collected for reporting. Such data can originate from conference- or meeting-based bookings made via housing authorities or meeting planners, or perhaps for travel that is booked and paid for by a client. Companies who allow open booking for all travel struggle to effectively locate travelers in a crisis, disclose any potential risks or alerts, or provide services to some travelers in the event of a crisis. Outside of suppliers with direct connections to open booking applications or parsers, even when your travelers are trained to e-mail those open booking itineraries to the required application for data capture, employers have no control over when they do this. Within a managed program (via most TMCs), all new bookings, modifications and cancelations are usually updated in the database in real time or close to it, providing employers with ample opportunities to mitigate risk in a number of ways when time is of the essence. Some well-known companies, offering travel-related solutions, claim that open bookings equate to more traveler choice and that their solutions can bridge the gap for any potentially missing data. When using an open booking application's itinerary data for security purposes, changes and cancelations can be a major issue. Some applications require user intervention to manually delete trips that have been canceled, or to resubmit trips for changes unless an update can be e-mailed or picked up by an e-mail scan.

Open booking data issues

Consider a situation where a trip is booked and ticketed via an airline website, the itinerary is e-mailed to the traveler, who either allows their inbox to be scanned or they forward the e-mail to the open booking application. Days later, the traveler needs to cancel that booking and rebook with another airline to travel with someone else from the company. The arrangements are made with the new airline, but the traveler forgets to delete the original trip in the open booking application. Now there are two trips in the system for the traveler. Imagine the confusion this could cause with employers if similar circumstances impacted multiple employees at the same time? A good managed travel program can still provide a variety of options, including easy methods of making reservations, yet still capture critical reservations data needed to effectively manage risk for business travelers. Trying to manage risk with a completely unmanaged booking process for the sake of open booking, even if it did offer more traveler choice, is not worth the risk, considering that in a crisis you have a higher likelihood of inaccurate data unlike if the traveler had booked via your managed program (via a contracted TMC working in conjunction with your TRM provider). Does that mean that managed program data is perfect? No, but if implemented properly, reservations data can be more tightly controlled.

Open Booking Case Study

On January 15, 2009, when US Airways flight 1529 went down in the Hudson River in New York City, a regional office for an employer received a phone call from an employee's relative who was hysterical, insisting that his family member was on that plane. The office in question contacted their TMC, but was unable to obtain any information on the traveler, so they then turned to the travel manager. By this time, the inquiring family member had intentions of coming into the office because he wanted "some answers," for which there were none at the time. Human Resources suggested that the relative contact the crisis response hotline, while dispatching security to the office in question to protect the facility and its personnel. Human Resources also advised the person to stay home for any communications, and for their safety, considering the person was so upset. It turns out that the traveler in question was on a legitimate business trip, but that the traveler had purchased the trip online (outside of the employer's managed program), with the traveler's personal credit card, and without using an open booking application for itinerary data capture. Because of this situation, it was difficult or nearly impossible to get helpful intelligence to the traveler or the traveler's family or to provide adequate resources and support, and had there been a death or severe bodily injury involved, the traveler wouldn't have been eligible for their corporate credit card's accidental death and dismemberment (AD&D) coverage.

Personal property and identity theft

Consider the personal losses of a business traveler whose hotel room was just broken into. What if as a result of such a theft, the traveler's identity was stolen? Will your company support the needs of the traveler to ensure that the traveler's assets and identity are preserved? The traveler wouldn't have been where the traveler was if it weren't for the business trip!

Identity theft has reached epidemic proportions globally, with plenty of statistics published by consumer advocacy groups and government agencies, such as the U.S. Federal Trade Commission. The U.S. Federal Trade Commission's *2014 Consumer Sentinel Network Data Book* listed identity theft as the top reported complaint by consumers for the 15th year in a row, with approximately 332,646 complaints. The act of traveling for business presents many opportunities for a traveler to be exposed to scam artists looking to steal the traveler's identity. While taking precautions may be inconvenient and time consuming, there are many things that business travelers can do to reduce their chances of having their personal information stolen, such as:

- Keep a copy of all account numbers and relative account information in a safe place that is separate from where debit and credit cards are kept.
- Put mail and newspaper delivery on hold. This can prevent mail theft or an indication that the person is away, which can lead to the person's home being robbed.
- Don't travel with a checkbook; use only credit cards and cash.
- Don't use debit cards as PINs (personal identification numbers) can be stored in some card reader devices and if the information is stolen, criminals could steal all of the cash available in the account(s) linked to the debit card.
- Notify credit card issuers prior to travel, especially if traveling internationally, so that they can authorize legitimate charges and notify the card holder promptly if activity on the account doesn't match their records.
- Use VPNs (virtual private networks) when using the Internet. If the traveler's company doesn't provide one, the traveler should purchase their own annual subscription.

What if your employee had prescription medicine that may have black market value and it got taken as well? Now, a theft has turned into a potential medical issue. Ask yourself the following:

- Some medicines cannot be refilled before their due date, and other medicines are not easily refilled before their due dates. Do you have the resources and support available globally $(24 \times 7 \times 365)$ to get those medicines replaced?
- Do you have the means to get the traveler replacement medicine before the traveler experiences any serious medical issues?
- What kind of medical support do you have available, particularly outside of the traveler's home country, should the traveler need immediate medical attention?

Having someone steal property from your hotel room or safe is bad enough, but when theft has happened, the event itself ends quickly. But if your computer is hacked, the problem could linger in many ways. Hotels are ideal places for business travelers to fall victim to hackers who not only may want access to some of your intellectual property, but to your identity as well. Referenced in subsequent chapters, there are tips about using hotel and public access Wi-Fi, if you must use them. However, by whatever means you access the Internet while on business travel (e.g., personal hotspot, or Wi-Fi with VPN, or other tools), try to not conduct any financial transactions or to log into financial-related websites while traveling. Losing personal passwords to e-mail accounts or other personal use websites can not only be financially damaging to the individual, but can occasionally be humiliating when private information is made public.

Mugging and pickpocketing

The most important thing to remember when faced with a mugging or pickpocketing incident is to not resist in the event of any confrontation and do not pursue assailants. Things can be replaced, but not your life or well-being. Your first priority should be to get away to a safe place, typically a business or well-lit public place with lots of people, where you can contact the authorities.

Traveling with prescription medicine

According to the United States CDC (Centers for Disease Control and Prevention), the percentage of adults aged 55–64, during the years 2009 to 2012:

- Percent of persons using 1–4 prescription drugs in the past 30 days: 55.6%
- Percent of persons using five or more prescription drugs in the past 30 days: 20.3%

Source: http://www.cdc.gov/nchs/data/hus/hus14.pdf#085

According to a 2013 report by CBS News Atlanta, approximately 7 in 10 Americans use prescription drugs.[17]

Consider that with such a large percentage of the working population taking prescription medications regularly, people taking medications need a basic understanding and awareness to always do their research prior to international travel about bringing the drugs with them into another country. In general, most countries allow up to a 30-day supply of legitimately prescribed medications, in their original bottle. More than 30 days of prescription medication on a traveler can be considered a violation of many country's laws, particularly when it comes to controlled substances, such as narcotic pain medication or psychotropic drugs. In some cases, it simply isn't enough to carry the original prescription bottles with medication in them; travelers may be required to carry additional documentation along with having filed advance approval forms to be in compliance with the jurisdiction in question. In particular, narcotics or psychotropic drugs must have extensive paperwork prepared by your doctor and submitted to the government of the country that you are visiting well in advance of travel, in order to process your paperwork for approval.

Employers must consider providing this kind of information to travelers with their pretrip briefings or risk reports, where applicable. The possibility of medicine being confiscated and/or criminal charges filed against someone for lack of approval to transport controlled substances into some countries is very real, and could cost someone their life if stranded on international travel without their medicine.

---

[17] CBS News, "Study Shows 70 Percent of Americans Take Prescription Drugs," June 20, 2013, http://www.cbsnews.com/news/study-shows-70-percent-of-americans-take-prescription-drugs/.

The adoption of this Convention is regarded as a milestone in the history of international drug control. The Single Convention codified all existing multilateral treaties on drug control and extended the existing control systems to include the cultivation of plants that were grown as the raw material of narcotic drugs. The principal objectives of the Convention are to limit the possession, use, trade in, distribution, import, export, manufacture, and production of drugs exclusively to medical and scientific purposes and to address drug trafficking through international cooperation to deter and discourage drug traffickers. The Convention also established the International Narcotics Control Board, merging the Permanent Central Board and the Drug Supervisory Board.

Article 36, Penal Provisions of Single Convention on Narcotic Drugs, 1961, as amended by the 1972 Protocol Amending the Single Convention on Narcotic Drugs, 1961, provides:

1. **a.** Subject to its constitutional limitations, each Party shall adopt such measures as will ensure that cultivation, production, manufacture, extraction, preparation, possession, offering, offering for sale, distribution, purchase, sale, delivery on any terms whatsoever, brokerage, dispatch, dispatch in transit, transport, importation and exportation of drugs contrary to the provisions of this Convention, and any other action which in the opinion of such Party may be contrary to the provisions of this Convention, shall be punishable offences when committed intentionally, and that serious offences shall be liable to adequate punishment particularly by imprisonment or other penalties of deprivation of liberty.
   b. Notwithstanding the preceding subparagraph, when abusers of drugs have committed such offences, the Parties may provide, either as an alternative to conviction or punishment or in addition to conviction or punishment, that such abusers shall undergo measures of treatment, education, after-care, rehabilitation and social reintegration in conformity with paragraph 1 of article 38.

According to the INCB (International Narcotics Control Board), at the time of this publishing, the following countries maintain strict regulations for travelers with restricted medications (see full list in the INCB "Yellow List" found at https://www.incb.org/documents/Narcotic-Drugs/Yellow_List/53rd_Edition/YL-53rd_edition_EN.pdf):

- Algeria
- Armenia
- Ascension Island
- Austria
- Azerbaijan
- Bahrain
- Barbados
- Belarus
- Belgium
- Belize
- Benin
- Bhutan
- Bosnia & Herzegovina
- Brazil
- Brunei Darussalam
- Bulgaria
- Burkina Faso
- Cameroon
- Canada
- Chad
- Chile
- China
- Colombia
- Costa Rica
- Cyprus
- Czech Republic
- Denmark
- Dominica

- Ecuador
- Eritrea
- Estonia
- Ethiopia
- Finland
- France
- French Polynesia
- Georgia
- Germany
- Ghana
- Häiti
- Hong Kong, China
- Hungary
- Iceland
- India
- Indonesia
- Ireland
- Israel
- Italy
- Japan
- Jordan
- Kazakhstan
- Kenya
- Korea, Republic of
- Kyrgyzstan
- Laos
- Latvia
- Lebanon
- Lithuania
- Luxembourg
- Macao, China
- Malaysia
- Maldives
- Malta
- Mauritius
- Mexico
- Micronesia
- Moldova
- Montenegro
- Montserrat
- Morocco
- Namibia
- Nauru
- Netherlands Antilles
- New Zealand
- Niger
- Oman
- Palau
- Panama
- Peru
- Poland
- Portugal
- Qatar
- Russian Federation
- Saint Lucia
- Senegal
- Seychelles
- Singapore
- Slovak Republic
- Slovenia
- Solomon Islands
- South Africa
- Spain
- Sri Lanka
- Sweden
- Switzerland
- Syrian Arab Republic
- Tajikistan
- Timor Leste
- Togo
- Tristan da Cunha
- Tunisia
- Turkey
- Uganda
- Ukraine
- United Arab Emirates
- United States of America
- Uzbekistan
- Vanuatu
- Zimbabwe

Source: International Narcotics Control Board, "Single Convention on Narcotic Drugs, 1961," http://www.incb.org/incb/en/narcotic-drugs/1961_Convention.html.

Measuring traveler wear and tear

Too much travel can burn many a road warrior out. The costs of this burnout are well known: lost productivity, increased safety risks, poor health, increased stress at work and home, unwillingness to travel, and, ultimately, increased attrition.

tClara, a travel data analytics firm, has developed a scoring system to track how much wear and tear each traveler accumulates from his or her travels. The goal is to predict which road warriors are at the highest risk of burnout, so that management can intervene in a timely manner.

The system uses a company's managed travel data to score a dozen factors found in each traveler's itineraries. Trip Friction[18] points are assigned to factors such as the length of the flight, the cabin, the number of connections and time zones crossed, the time and day of week of each flight, etc. This allows for traveler-specific and company-specific benchmarking, which in turn helps senior executives to influence travel policy, procurement strategy, and traveler behavior to optimize a managed travel program.

Traveler friction versus travel policies

Push travelers through too many pain points, and the traveler may soon find reasons to not take the next trip. For example, think about flying coach from Chicago to Singapore, or taking a short haul connection for a lower fare. Tighten the travel policy too much, and you could have recruiting and retention problems, which could have serious cost or business implications. Companies shouldn't focus solely on minimizing the transaction cost of their trips; instead, they should focus on minimizing the total cost of traveling. That's the sum of the trip's transaction cost plus the cost of traveler friction (the black curve in the figure below) or the "Total Cost Paradigm."

The total cost of travel paradigm formalizes what buyers do intuitively

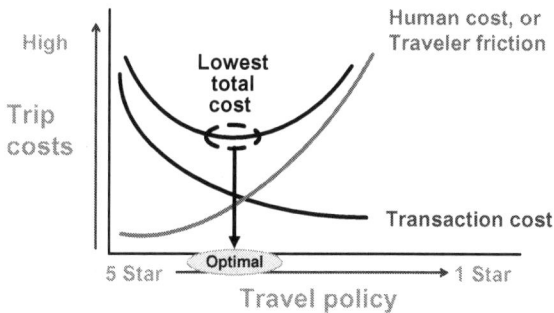

High

Trip costs

Lowest total cost

Human cost, or Traveler friction

Transaction cost

5 Star — Optimal — 1 Star

Travel policy

The total cost of travel paradigm formalizes what buyers do intuitively.

tClara's CEO, Scott Gillespie, states the following:

Travel programs depend on travel policies for savings. The tighter or tougher the travel policies, the cheaper the transaction costs, such as airfares, hotel stays, and car rentals. Our industry understands the shape of the blue curve in the chart above. However, as you increase the toughness of that travel policy, you create costs as well. Costs we will call "traveler friction," as shown by the red curve.

---

[18] "Trip Friction" is a registered trademark of tClara.

The total cost of travel paradigm

Goal: Minimize the total cost of travel, including **quantifiable** traveler wear and tear

Principles

1. **Budget owners** set <u>traveler</u>-related goals; e.g., recruiting, retention, health and safety, etc.
2. **Procurement and HR** agree on metrics for tracking traveler wear and tear costs
3. Travel managers provide **travel strategy plans** to help achieve the traveler-related goals
4. Monitor goals and **total travel costs**, and adjust travel strategies accordingly

The total cost of travel paradigm.

Figure below reflects a precedent acceptance of total cost concepts, along with examples of how the travel industry is beginning to acknowledge the HR costs of traveler "wear and tear."

Good news: procurement understands the total cost concept

- U.S. auto industry adopted total cost of quality in the 80s
- IT and Procurement functions adopted total cost of ownership in the 90s

The travel industry is starting to quantify the HR costs of traveler wear and tear

- CWT's traveler stress index
- BP's focus on traveler safety
- Accenture's 3/4/5 travel policy
- tClara's trip friction scoring method

Good news.

From a TRM perspective, firms should monitor traveler-related metrics such as these:

- Work days lost;
- Attrition rate;
- Time to fill travel-intensive jobs; and
- Accumulated Trip Friction points and benchmarks.

These metrics will give management a view about how hard their most frequent travelers are traveling, and whether or not the travel policies should be adjusted.

To put Trip Friction into perspective, tClara provides two trip examples (refer to the figure below) showing a low level of Trip Friction in "Trip A" versus a higher level in "Trip B."

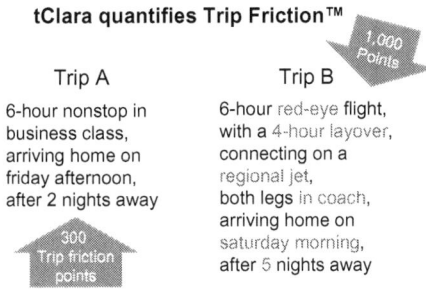

tClara quantifies Trip Friction™

1,000 Points

Trip A	Trip B
6-hour nonstop in business class, arriving home on friday afternoon, after 2 nights away	6-hour red-eye flight, with a 4-hour layover, connecting on a regional jet, both legs in coach, arriving home on saturday morning, after 5 nights away

300 Trip friction points

tClara quantifies Trip Friction.

According to tClara (refer to the figure below), their data shows a correlation between Trip Friction and higher numbers of road warrior or frequent traveler turnover.

Trip friction is clearly correlated with higher road warrior turnover

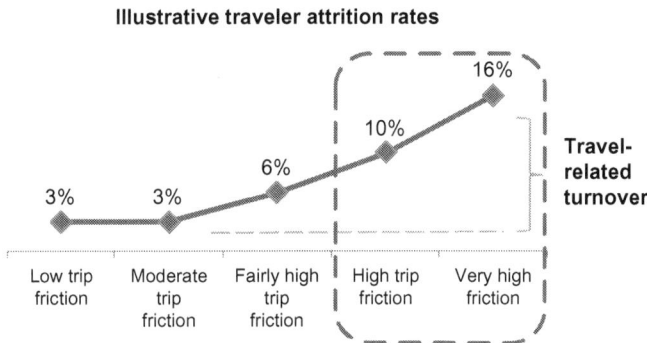

Illustrative traveler attrition rates

Trip Friction is clearly correlated with higher road warrior turnover.

While strong travel policies under managed corporate travel programs are critical to successful TRM (versus unmanaged, open booking allowances), there is a delicate balance between cost savings, safety, traveler satisfaction, and, very importantly, business continuity. Trip friction and traveler friction are good examples of the link between TRM and operational risk management (see Chapter 9), which shows how losses of productivity or employees managed under the guise of TRM can impact company production and/or success.

Personal well-being and stress

Personal well-being of travelers might be the most surprising of topics for consideration, but it certainly is relevant in context with TRM programs today. Believe it or not, employers must be as cognizant of their employees' or contractor's mental well-being as of their physical safety. Stressed out, tired, or even unhappy employees can represent lower productivity and a higher threat of risk.

From something as simple as knowingly requiring someone to work in a stressful environment without trying to make it better, or just working them to excess, can cause an employee to suffer various forms of posttraumatic stress or depression. However, it can also be as extreme as requiring employees to work in a stressful situation without being properly trained or counseled, as was the case with some flight attendants who may have been forced to immediately fly again out of New York after witnessing the 9/11 attacks, when the commercial flights began operating again, without consideration of stress or trauma, proper treatment, and counseling.

To the extent that employers monitor and evaluate the physical safety of employees or contractors in the workplace, they must now take notice of the level of employee/contractor stress and contribute to overall happiness. It turns out that employees with high states of well-being have lower health care costs.[19] It's unfortunate that employers must usually see a financial benefit associated with such things before implementing them, but in addition to health care costs, if people are happier and healthier, it stands to reason that they are also more productive.

The CWT Solutions Group conducted a study to shed light on the hidden costs of business travel caused by travel-related stress. Their aim was to understand and measure how and to what extent traveler stress accumulates during regular business trips. They defined a methodology and a set of key performance indicators (KPIs) to estimate the impact that this travel-induced stress has on an organization (see "The Carlson Wagonlit Travel Solutions Group Study").

The Carlson Wagonlit Travel Solutions Group Study

The scope of the study includes data from 15 million business trips booked and recorded by Carlson Wagonlit Travel (CWT) over a 1-year period. They followed a *divide-and-conquer* approach: each trip was conceptually broken down into 22 potentially stressful activities covering pretrip, during trip (transportation- and destination-related elements), and posttrip. Associated stress was measured based on the duration and the perceived stress intensity for each activity. In essence, each of the 22 steps of the trip was viewed as having two components: *stress-free* time and *lost time*.

To quantify the effects of stress, we introduced the following KPIs [key performance indicators]:

- The maximum possible lost time per trip
- The actual lost time per trip (and its financial equivalent)
- The Travel Stress Index, defined as the ratio of the above quantities

[19] Susan Sorenson, "Lower Your Health Costs While Boosting Performance," *Business Journal* (Gallup), September 19, 2013, http://businessjournal.gallup.com/content/164420/lower-health-costs-boosting-performance.aspx.

The Travel Stress Index (TSI) across all trips booked through CWT is 39%. Our results show that the actual lost time is 6.9 hours per trip, on average. The largest contributions to this lost time arise from flying economy class on medium and long-haul flights (2.1 hours) and getting to the airport/train station (1.1 hours). The financial equivalent of this 6.9 hours is US$ 662.

The lost time greatly depends on the type of trip taken: an increase in the transportation time typically generates an increase in the lost time. The average actual lost time values by trip type are:

- 5.2 hours for domestic trips
- 5.6 hours for continental trips
- 15.6 for intercontinental trips

Finally, the study indicates that the impact of stress can be reduced, but not entirely eliminated. They analyzed the TSI on a client-by-client basis and found out that companies can expect to control, on average, 32 percent of the actual lost time.

TSI survey data

In a previous publication [Ref. 1], CWT Solutions Group presented the perceived stress reported for 33 activities related to a typical business trip. The current study incorporates 22 of these factors (Table 1.1), including nine of the

Table 1.1 **Stress-trigger ranking by perceived strength**

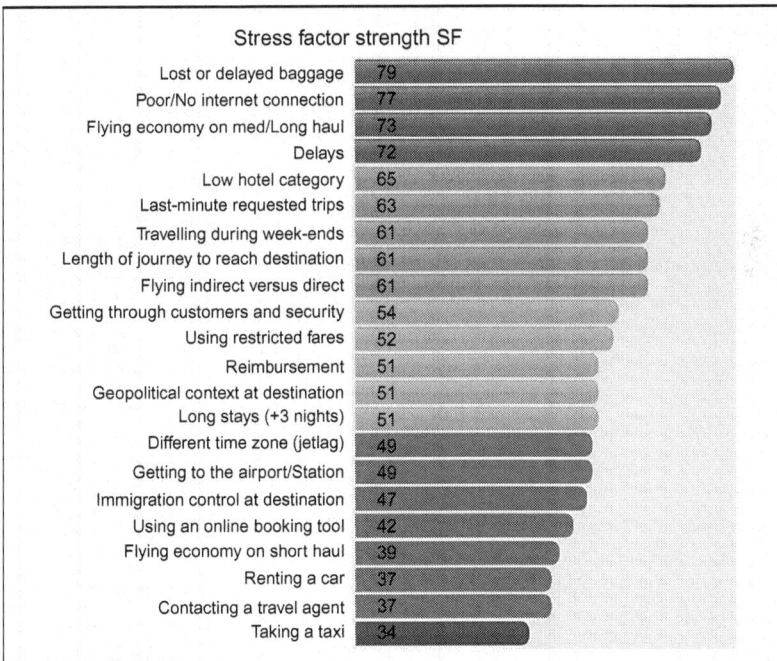

Stress factor strength SF	
Lost or delayed baggage	79
Poor/No internet connection	77
Flying economy on med/Long haul	73
Delays	72
Low hotel category	65
Last-minute requested trips	63
Travelling during week-ends	61
Length of journey to reach destination	61
Flying indirect versus direct	61
Getting through customers and security	54
Using restricted fares	52
Reimbursement	51
Geopolitical context at destination	51
Long stays (+3 nights)	51
Different time zone (jetlag)	49
Getting to the airport/Station	49
Immigration control at destination	47
Using an online booking tool	42
Flying economy on short haul	39
Renting a car	37
Contacting a travel agent	37
Taking a taxi	34

Adapted from [Ref. 1].
Source: CWT Solutions Group, Stress Triggers for Business Travelers, Traveler Survey Analysis (2012).

top 12—those with scores above 60/100. The remaining 11 factors are either challenging to quantify (e.g., "eating healthily at destination") or require certain data that was not available at this time.

Several stress factors, such as flight delays, mishandled baggage, and traveling to a high-risk destination, require the usage of *external data*. References [6], [7], and [8], respectively, are used for these purposes.

CWT Solutions Group References as noted in "Travel Stress Index—The Hidden Costs of Business Travel:" [20]

[1] CWT Solutions Group, "Stress Triggers for Business Travel" (2012). This research is available for download at: http://www.cwt-solutions-group.com/publications-and-media-centre/research-and-view-points/stress-triggers-for-business-travelers.html.
[6] FlightStats, Inc. is a leading publisher of flight information to travelers and businesses around the world (www.flightstats.com).
[7] Baggage Report 2012, SITA (www.sita.aero).
[8] *iJET* (www.ijet.com) is an intelligence-driven provider of operational risk management solutions, working with more than 500 multinational corporations and government organizations.

[20] http://www.cwt-solutions-group.com/export/sites/cwt/cwtsg/.content/files/CW-cwt-travel-stress-index-hidden-costs.pdf

Repatriation of mortal remains

Unfortunately, people sometimes die while away from home on business. Making arrangements to transport their remains across international borders can be complicated and expensive, as legislation and protocols vary greatly from country to country, as do suppliers who will provide such services. Don't assume that your TMC will or can handle this for you. Usually these situations are handled by medical emergency or insurance providers.

The following items should be covered in repatriation of mortal remains insurance:

• If passing takes place outside of a medical facility, adequate transportation (ambulance, airplane, or helicopter) equipped with proper storage and handling capabilities for the body during transport to the closest appropriate medical facility prior to international transport.
• Treatment costs incurred (including embalming).
• Legally approved container for shipment of the remains.
• Transportation costs for the deceased and an accompanying adult to the country of residence.
• Cremation if legally required (conditional).

Other coverage may be included for things such as hotel accommodations pre- or posttreatment prior to the passing of the insured, but coverage will vary widely between providers. Under such stressful circumstances, it is very important for the insured's family to understand the claims process and coverage, such as will payment be provided directly to suppliers for services as needed, or will prepayment be required by the family or loved ones, only to request reimbursement later? If it can be avoided, such understanding can reduce stress associated with paperwork, authorizations, and payment.

Theft of intellectual property

According to the Cornell University Law School, in general terms, intellectual property is any product of the human intellect that the law protects from unauthorized use by others. The ownership of intellectual property inherently creates a limited monopoly in the protected property. Intellectual property is traditionally comprised of four categories: patent, copyright, trademark, and trade secrets. In summary, if you are in business, you likely have some intellectual property to protect. It could be an idea, or simply a process that you use, which gives you a competitive edge.

Most people think of a stolen laptop or mobile phone when they think of vehicles for stolen intellectual property, but a far more common vehicle is a flash drive, which most business travelers carry with them today on business trips and aren't monitored or regulated in the same manner as phones, computers, or tablets. Companies should either limit the use of flash drives to those drives that have some level of FIPS (U.S. Federal Information Processing Standard) to encrypt the data and/or destroy the data should the drive be tampered with physically in an attempt to access its contents.

Information on current FIPS standards (FIPS 140-2) and announcements regarding the upcoming FIPS 140-4 standard, can be found by visiting http://csrc.nist.gov/groups/STM/cmvp/standards.html#05.

Traveling light

Many companies have policies specific to certain countries whereby, when travelers intend to visit the countries in question, the travelers either cannot take laptops or standard mobile devices with them, or the travelers must take "clean machines" or hardware designed for travel specifically to countries with high numbers of intellectual property theft. Some of this hardware may have special configuration or software to add layers of protection, in addition to not storing important files locally (i.e., cloud computing), or transportation of valuable files is done via one-time-use USB flash drives.

International protections for intellectual property rights

Because there are times when identifying intellectual property thieves can be nearly impossible, one might not have the opportunity to take advantage of any legislation or treaties. However, it is good to know that programs are developing and in place to try and protect intellectual property owners, such as the TRIPS (Trade Related Aspects of Intellectual Property Rights) Agreement from the WTO (World Trade Organization). TRIPS was designed to set some standards for how intellectual property rights are protected around the world under common international rules. These trade rules are seen as a way to provide more predictability and order, and a system for dispute resolution, providing a minimum level of protection for all WTO member governments.

For more details on the TRIPS Agreement, see https://www.wto.org/english/thewto_e/whatis_e/tif_e/agrm7_e.htm.

HIV-positive travelers and expatriates

As of May 2015, 36 countries place various forms of restrictions for the entry, stay, and/or residence of people who are HIV-positive.[21]

In 2009, the United States removed its entry restrictions for people living with HIV, which received considerable media coverage and is believed to have had an influence on many another country's legislation on the matter, as the number of countries with such restrictions has declined from 59 in 2008 to 36 in 2015.

Restrictions vary from country to country, but are broken down into the following categories:[22]

- Countries without restrictions
- Countries that ban entry
- Short-term stay restrictions—less than 90 days typically
- Long-term stay restrictions—more than 90 days typically
- Countries with unclear laws or practices
- Countries without information
- Countries deporting people with HIV

Detailed lists of countries with corresponding information on legislation can be found at The Global Database on HIV-Specific Travel & Residence Restrictions.

Reminder: Although this text provides various reference materials found on the Internet, there is no substitute for or comparison to the quality of medical and security intelligence created, monitored, and provided by qualified risk intelligence providers, which are at the core of employer-managed TRM programs. One specific reason for the importance of risk intelligence providers is because guidelines, laws and requirements regularly change.

What is surprising to realize is that some of the countries from which an HIV-positive traveler could be deported if the traveler's HIV status were known, are countries that are common destinations for many business travelers today. Imagine a security check that uncovers prescription HIV treatment medication in a country where there are entry restrictions? This is a difficult position for employers because of the privacy concerns of employees or travelers and their medical records, which are not typically the kinds of records or information that a person shares with employers.

---

[21] UNAIDS, "Infographic: Welcome (Not): Before and After," May 26, 2015, http://www.unaids.org/en/resources/infographics/20150227_evolution_travel_restrictions.

[22] The Global Database on HIV-Specific Travel & Residence Restrictions, http://www.hivtravel.org/Default.aspx?pageId=152.

However, just as with prescription medications that people can travel with, employers need to provide appropriate training and information to travelers going to places where HIV concerns may be an issue. While adding this kind of information on top of standard risk and policy disclosures may be an extensive and painfully large amount of information to read and understand prior to travel, employers have a duty to provide it, and travelers have a duty to understand it and act accordingly if one or more of any disclosed travel restrictions apply to them.

In some of the more strict countries with legislation that allows deportation of HIV-positive travelers, deportation often doesn't apply to travelers connecting or in transit only. However, employers and travelers have to decide whether or not they want to take such a chance. Some countries require medical exams for those who intend to stay longer than 30 days, and if HIV is discovered, doctors are required to report it to the government, and the law will be administered relative to the country in question.

Assessing your current approach to travel risk management

Unfortunately, even now with constant civil unrest, natural disasters, pandemics, and various other hazards, it seems that many companies are still simply reactive instead of proactive when it comes to mitigating or dealing with a crisis or traveler-related emergencies. Since 9/11 an increasing number of progressive companies have developed truly forward-thinking programs that attempt to get ahead of the risks before they happen; many companies, however, remain unprepared. With each jurisdiction around the world having its own laws and subjective views on what is enough or appropriate in terms of providing duty of care, as well as building and maintaining a travel risk management (TRM) program, employers need a standardized approach for measuring their effectiveness in this area, and for comparing their efforts to the efforts of other similar companies.

From a fundamental perspective, companies need to understand that TRM and solutions for managing their duty of care responsibilities consist of many individual components. Contrary to what some technology providers in the market tend to believe, TRM isn't found holistically in a piece of software, or a single insurance policy. The best and most recognized standard for assessing a company's current ability and effectiveness in managing travel related risks is a program designed and trademarked by iJET International, called TRM3, which stands for "Travel Risk Management Maturity Model." More than an assessment, TRM3 is a framework and discipline for continuous process improvement of a company's TRM program.

The model defines "business travel" as any time an employee represents his or her organization away from home, either domestically or internationally. This can range from a drive to a facility in another city or abroad, or on to expatriate assignees (long-term assignments) that face more potential threats than do average travelers and therefore have significantly higher risk profiles. Among the types of threats and hazards travelers face, can range from a trip to the airport, to petty crime and terrorism. However, iJET claims that threats are no longer risks if made irrelevant or properly mitigated. To this end, it defines "risk" as:

$$\text{Threat} - \text{Mitigation} = \text{Risk}$$

However, according to iJET's TRM3 white paper, risk management is not a formula. It's a process, made up of the following steps:

1. Identify relevant threats (threat environment).
2. Evaluate threats in relation to a traveler's profile (relevance).
3. Set an acceptable level of risk for the organization and employee (risk appetite).
4. Implement mitigation strategies that reduce threats to an acceptable risk level (mitigation).
5. Monitor for any changes in threats or a breakdown in the mitigat.ion strategy (monitor).
6. Respond to an incident when it occurs (respond and recover)

Defining travel risk management

TRM means a lot more than reacting quickly and efficiently to events as they unfold. In fact, the only reactive component of a sound TRM program is incident response. All other components (policies and procedures; training; 24 × 7 monitoring, including traveler tracking; feedback) must be planned, implemented, and—importantly—practiced before travel begins. By establishing a continuous process loop and training employees to follow it, all manner of risk (not only travel-related risk) can be significantly mitigated.

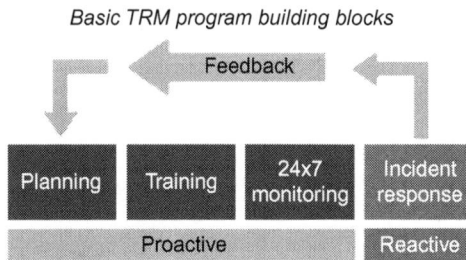

Basic TRM program building blocks.

Understanding how these elements correlate to one another validates the point that a proactive TRM program isn't a standalone project. It is a 24 × 7 process.

Common mistakes when beginning to assess a company's risks include:

- Assuming that traveler tracking alone constitutes TRM. (This is answered by addressing each of the TRM3 key process areas.)
- Claiming that the company's travelers don't travel to high-risk destinations. (This is irrelevant and subjective. Risks can and do change frequently, even for normally low-risk locations.)
- Claiming that because the company's travel is primarily domestic and not international, the company doesn't need a TRM strategy. (A crisis or serious incident can happen anywhere, anytime, possibly only affecting one traveler. Lack of preparation can prove to be negligence in a court of law.)

Using the TRM3 model, company stakeholders are directly engaged by travel risk and security professionals, with interviews and various forms of discovery, to produce an in-depth analysis and scorecard, which in effect is a well-rounded business plan for where to start in building or adapting one's approach to TRM. The final result also acts as a well-documented business case for continued investment in the plan, most especially because the data is benchmarked against other companies of similar size and industry globally. Short of a full-scale professional services engagement, there are trained specialists in the industry who can assist companies with a light version of the process, using a scorecard template document, which has been adopted and endorsed by the Global Business Travel Association (GBTA).

According to iJET's TRM3 white paper (http://info.ijet.com/resources/whitepaper), the following is an outline of the five key components or assessment areas of a proactive TRM program.

1. **Planning:** An organization needs to define its overall TRM strategy, linking TRM policies to key organizational goals. This means determining how the TRM program will integrate with local crisis management plans (CMPs) and business emergency plans (BEPs). The main objective in this phase is to plan now, so that the organization can do more than simply react later. There are a wide range of questions to consider during the planning phase, for example: What is the company's response if an employee is kidnapped or killed? How will the company evacuate employees in an emergency? What if an employee becomes seriously ill while traveling? These are all incident types that should be addressed during the planning phase.

2. **Training:** The purpose of training is to develop employees' skills and knowledge so they can perform their roles effectively and efficiently. In this paper (iJET's TRM3 white paper), we define three specific areas of training that should be addressed: traveler training, travel advisor training, and crisis management team training.

3. **24 × 7 Monitoring:** Systems and staff need to be put in place to provide real-time monitoring of potential threats to travelers worldwide. Through automated itinerary and assignment monitoring, the iJET TRM program notifies clients of any high-risk trips or assignments. When a threat is determined, getting this relevant information and possible mitigation strategies into the hands of the traveler or an advisor is critical. With advanced notification, many problems can be avoided.

4. **Incident Response:** Employees need to have someone to contact day or night for help in cases of emergencies. An optimized TRM program would be integrated into an organization's overall operating risk management (ORM) program. The resulting ORM program would include a single emergency hotline service for any issue or emergency impacting travel, facilities and/or supply chains.

   Understanding that no single internal resource or response vendor can handle every incident type in every location around the world, iJET has developed a "Command Center" infrastructure and incident management system to coordinate multidisciplinary response from multiple vendors. This system is customized for every organization and performed under the direction of the organization's crisis management team (CMT) or incident management team (IMT). For the organization, this solution enables global awareness of any incident that may impact people or operations. For the employee, it provides one number, worldwide, for any problem.

   Recent events, such as the violent civil unrest across the Middle East and North Africa, as well as the devastating natural disasters in Japan, underscore the need for an integrated incident management approach that extends beyond travel. During and following these events, corporate travel departments learned, again, that they are only part of a robust incident response that requires integration with a number of departments across their organization. Organizations with well-developed, integrated incident management plans, processes, and procedures fare significantly better than those with underdeveloped and/or disjointed programs.

5. **Feedback:** Following any incident, it is important to have an after action review (AAR). The AAR asks: "Could the problem have been prevented in the first place?" And if not: "Could the incident have been handled more effectively?" If the answer to either of these questions is "Yes," then a modification of existing policies, plans, procedures, or mitigation

strategies is required. This feedback process could be extended to a short survey after each trip. This would provide valuable information about the organization's travel program and capture any issues or concerns employees might have. Risk management must be an ongoing process under continuous improvement.

What is an employer's "duty to disclose"

In Chapter 1, we discussed the concept of "duty of care," and further to that concept which requires "reasonable best efforts," is something called "duty to disclose," which is a concept that is based upon providing timely information and intelligence before someone travels or is exposed to risk, so that an informed decision can be made as to whether or not to go on the trip and/or to be prepared when exposed to the potential risk. What organizations don't want to have to defend in litigation is a traveler with damages who says, "I was never advised of these risks," and the employer not having any proof that it disclosed said risks before the employee's exposure to the incident. That is why a duty to disclose is so important, because employers have no defense without such proof. This proof must be automated and not sent manually by administrative assistants or travel arrangers, who subscribe to some type of alert feed. Such manual efforts are an extremely shortsighted attempt to cover an employer's duty to disclose in an attempt to save money, and it is a dangerous game that is subject to human error by not sending the appropriate information, which may require more than just an alert, or not sending the right information relevant to a particular trip. While there may be cost-effective or competitive means to achieve a good, automated method for a duty to disclose, the words "cheap" and "safety" should not be used in relation to one another. Smart value is an altogether different concept.

Push versus pull disclosures

Disclosures can come in the form of multiple communications from alerts to in-depth risk reports developed specifically about a destination where a traveler plans to visit. A "push" disclosure simply means that you are "pushing" the information to the recipient, typically via e-mail, SMS text message, or a mobile application "push" message. A "pull" disclosure means that a company has made information available to travelers (such as a database or website), but it is up to the travelers to go to the source of information, locate the appropriate information, and make informed decisions. Pull disclosures should only be used as a supplement to push disclosures, which should be the primary method of informing travelers of potential risks.

Timely push disclosures are most definitely a more effective method for distribution of information, and are easier to prove and use during conflict or litigation.

In addition to potential liability associated with a crisis or critical incident, when duty to disclose is implemented effectively, perhaps in conjunction with a travel authorization process, time, labor, and financial losses may be avoided if smart decisions can be made in advance. For example, an employee needs to be in London for a critical business meeting during a time when a planned ground transportation strike is scheduled to take place. Going through with the trip could place the employee in London when the potential customers can't get to the meeting place because of the stoppage, thus canceling the meeting, costing the employers thousands of dollars in travel expenses for the first trip and potentially a second one!

Travel risk management maturity model (TRM3) overview

TRM3 is both an assessment process, a framework, and a continuous process improvement discipline to create and refine a company's processes with respect to its TRM program. At a macro level, companies should work towards creating many custom processes and protocols within their TRM program, such as one example provided below.

Travel risk management (TRM)
program overview

Pre-trip intelligence
Safe return home
Book trip
24×7 global assistance
Automated risk assessment
ASSOCIATE
Report trouble
Emailed trip briefs & alerts
Personal web site

Sample TRM program overview from iJET's TRM3 white paper.

The TRM3 maturity model provides.

- A starting point;
- Benefit of community precedent or experience;
- Common language, vision, or goals;
- A prioritized framework; and
- An outline for organizational improvement.

The TRM3 model has been designed to address 10 key process areas (KPAs). When evaluating an organization's maturity, one must use the lowest KPA score as the overall rating for the organization. For example, an organization's Policy/Procedures KPA may be at a Level 4 and the Risk Mitigation KPA may be at Level 2. These ratings would provide a TRM3 rating result of Level 2.

MATURITY LEVEL	DESCRIPTION
Level 1 – reactive	The TRM process is characterized as ad hoc and can be chaotic in the event of an incident or emergency. Few policies, procedures, or processes are defined and success depends on individual effort and available resources. This reflects a program that does little to proactively manage travel risk, with staff simply reacting to events as they occur.
Level 2 – defined	Here, an organization has defined and documented key safety and security protocols within its travel program, with a particular focus on incident response. However, it is missing or inconsistently provides risk disclosure, mitigation, monitoring and the other elements of a proactive program. There is also a heavy reliance on manual processes, which are subject to human error. Policies and processes are not consistently applied. This is typically the result of a lack of training, program communications, supporting data management systems and integration into the day-to-day travel management program.
Level 3 – proactive	The organization has incorporated some form of employee training, risk disclosure and notification process as part of a formalized TRM program. Automated systems have been introduced to support the program. The TRM process for both management and travelers is being consistently executed within the travel department. A risk assessment is performed for each trip. This assessment results in proper management notification and risk disclosure to both the traveler and the organization. Travelers are aware of their responsibilities, travel policies and safety practices through a consistent training program. An emergency assistance and response structure is in place to support the traveler and/or organization during any type of incident (security, medical, transportation, etc.). However, systems are not applied consistently across the organization and there is no effort to measure the effectiveness of the program.
Level 4 – managed	The organization that has adopted all KPAs organization-wide, with appropriate systems in place to support the program across regions and business units. Processes are consistently applied and executed. Detailed metrics surrounding the TRM program are collected and reviewed. At this level, the TRM program is embraced and supported by management and across the organization. The emergency assistance and reponse structure is unified within the overall organization's crisis and emergency management program.
Level 5 – optimized	This is the highest level of program maturity. At this level, the travel risk program is integrated throughout the organization and is well understood by management and employees, with automated compliance monitoring. Metrics and lessons learned are collected and used to continuously improve the program. At this level, continuous process improvement is enabled by quantitative feedback and lessons learned. There is a program to pilot innovated ideas and technologies. Process changes and/or new technologies are adopted into the overall program.

TRM3 maturity levels explained from iJET's TRM3 white paper.
Source: iJET, "White Papers: Travel Risk Management & Maturity Model (TRM3),"
http://info.ijet.com/resources/whitepaper.

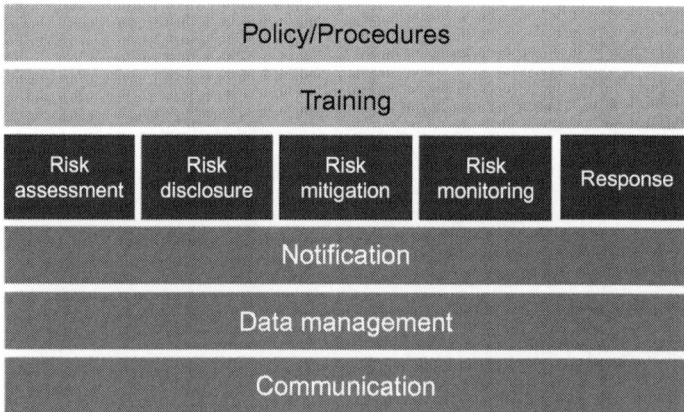

TRM3 KPAs from iJET's TRM3 white paper.
Source: iJET, "White Papers: Travel Risk Management & Maturity Model (TRM3),"
http://info.ijet.com/resources/whitepaper.

Policy/procedures

The purpose of the Policy/Procedures area is to focus attention on the process of developing and maintaining policies and procedures in support of TRM.

To reach maturity in the Policy/Procedures KPA, organizations must ensure that policies are well defined and documented, broadly integrated within organizational risk policies, implemented within the overall travel process, measured and monitored for compliance, and supported by a continuous improvement process. Reaching maturity in the Policy/Procedures KPA means developing, implementing and maintaining policies and procedures that are at a minimum of parity with other similar companies, yet improve year over year.

Typical risk management policies include the use of private aviation, limitations on the number and type of employees on the same flight, prohibited travel destinations, approval process to high-risk destinations, etc., and all TRM policies and procedures should address several important areas, including vetting of vendors and suppliers, business continuity of travel operations and call centers, and incident and emergency response protocols. These policies and procedures should be developed as part of the other KPAs defined below.

Key considerations for employers when addressing policies and procedures as a part of their TRM strategy include, but are not limited to the following:

- Defining processes, parameters and best practices around aspects of travel in context with policies and the law
- Cultural and/or regulatory requirements on the country, regional, and global levels
- Process for regular review and updates of policies and procedures
  - Includes measurable stakeholder deliverables
- Benchmarking policies and procedures at a minimum annually
- Definitions of the program, its purpose, and key terms and definitions
- Reservations and trip data capture requirements in support of TRM program (allowance or prohibiting of open bookings, and approved technology for use in data capture if allowed)
- Specific policies and procedures that individually address health, safety, and security
  - Pretrip risk-mitigation processes and strategies
    - Travel alerts or warnings
    - Crisis communications
    - Training
      - How and when to partake in pretrip, traveler safety training
    - Equipment issuance (clean hardware)
    - Insurance coverage for:
      - Medical services and evacuations
      - Security services and evacuations
      - Car rentals
      - Meetings and events
      - Group activities and team building
      - Suppliers
    - Selection and process for usage of safety approved suppliers
    - Allowance or prohibiting of sharing economy supplier usage
- Pretrip requirements for travel to high-risk destinations
- Pretrip medical consultations and requirements (i.e., vaccinations)

- Protection of intellectual property
- Social media use guidelines
- Personal trip extensions to business trips
- Safety requirements and required processes for meetings and events
- Antibribery legislation
- Crisis response and support training (what to do?)
- Kidnapping and hostage survival procedures
  - Proof of life requirements
- Crisis response support procedures and services
  - Medical support
  - Security support
- Compliance of the policies and penalties for non-compliance
- Policies regarding training (required and optional)
- Process improvement and feedback

Training

The purpose of training is to develop employees' skills and knowledge so they can perform their roles effectively and efficiently with minimal or no interruptions because of risk, safety, security, or medical-related issues.

There are three specific areas of training that should be addressed: traveler training, travel advisor training, and crisis management training.

Traveler training

This training covers all the essential issues from pretrip planning, to skills on the road, to decompressing when a traveler gets home. Some typical courses, either required or optionally offered by employers could include:

- Security awareness training, e.g., identifying potential risks or hostile environments, travel-related medical restrictions (conditions or medicines)
- Destination research and planning (including what to take or not to take)
- Secure ground transportation best practices
- Hotel safety training
- Crisis preparation and response
- Cyber security, e.g., avoiding intellectual and personal property theft
  - Social media policy and safety training
- "On the move" or "in transit" safety training (airports, car rentals, ground transport, hotels)
- Dealing with local authorities
- Defensive driving
- Self-defense training
- Surveillance detection
- Hostage survival training
- Natural disaster survival training
- Minority traveler training (race, sex, sexual orientation, religious affiliation)
- Custom expatriate training

Specialized traveler training may also be a consideration by different industry verticals, such as maritime or oil and gas industries. In particular, because of their unique laws and operational/occupational safety requirements, specific to their businesses, such as:

- Maritime customs, laws, and rules
- Safety of Life at Sea Convention[1]
- Emergency action drills (EADs)
- International Ship and Port Facility Security Code (ISPS)[2]
- Detention, treatment, search, and seizure of hostile personnel
- Hazardous materials/explosives awareness training
- Water survival and medical/rescue training

Travel advisor training (travel agents or counselors, not corporate administrators or "bookers")

Training for corporate travel agents with regards to traveler safety and security will vary to a degree based upon how much the travel management company gets involved in active TRM program management with their clients. For the record, travel management companies (TMCs) should not be in the business of taking on tactical responsibilities that involve fulfilling the needs of an employer's risk policies, such as verbal consultations related to safety, manual efforts, and research to provide or relay safety and security-related information to travelers. While TMCs may be a conduit for technology or security intelligence through programs offered to their clients, they should never be put in a position to be asked to personally provide safety or security advice, or bear the responsibility of a crisis response or support protocol task. These should be the responsibility of the client/employer. Standards for risk tolerance, policies, and procedures always rest with the employer. For example, even if and when a TMC provides its version of a crisis communication, each client should maintain its own standards for crisis communications for when they pertain to business travel and when to communicate to travelers and stakeholders, which may vary from the standard used by the TMC. Additionally, another good example of a responsibility that should not be undertaken by the TMC or travel advisor is, to research, provide risk related advice or intelligence disclosures via manual efforts as part of their standard reservations and ticketing processes, which ideally should be automated and therefore not subject to human error. Providing this kind of information to travelers is called a "risk disclosure" and is one of the KPAs in the TRM3 risk assessment model. Risk disclosures should be automated in conjunction with the trip data provided by the company's TMC combined with the intelligence and automation provided by the company's third-party risk intelligence provider or by the

[1] See United Nations, "Treaties and international agreements registered on 30 June 1980 No. 18961: Multilateral: International Convention for the Safety of Life at Sea, 1974, " https://treaties.un.org/doc/Publication/UNTS/Volume%201184/volume-1184-I-18961-English.pdf.
[2] See Maritime and Port Authority of Singapore (MPA), "ISPS (International Ship and Port Facility Security) Code", http://www.mpa.gov.sg/sites/port_and_shipping/port/port_security/isps_code.page.

employer's in-house security resources. When it is the latter, it is then the responsibility of in-house security resources to assess each trip and disclose those potential risks to travelers. What advisors should understand is exactly where they can help, and where they should not.

Examples of what advisors should **not do** or provide to clients include:

- Conduct individual searches of risk intelligence per trip, per client to either verbally provide or manually document and provide to clients.
- Provide any safety, security, or medical advice verbally or via manual e-mail distribution that is offered personally, or by repeating third-party information. Travel advisors or agents should never be a primary conduit of manual safety and security information for transient business travel, but instead should support the program that ensures automatic distribution of materials relevant to the traveler's trip.
- Creating or manually distributing any risk related documentation electronically or otherwise.

Examples of what advisors might consider providing limited support for to clients include:

- Thorough understanding of their role or place in supporting a client's TRM program via a documented process in support of TRM technology solutions and reporting.
- Support operational processes that trigger automated distribution of risk intelligence or processes, using technology and third-party risk intelligence (focus on the technology-based automation, not manual processes).
- Remind travelers of defined policies, processes, or resources where appropriate, reminding them of where they can access formal documentation or support from their employers, should they have specific questions or need safety and security related support.
- Possibly transfer calls to designated TRM resources when applicable, as part of a defined TRM support process for advisors, such as to a security officer, crisis response call center, or corporate crisis response team member.
- Arrange for and confirm client-approved safety and security services in conjunction with corporate travel arrangements (customer contracted and vetted secure ground transportation, personal security services, secure hotel accommodations, etc.).

Travel risk management employer stakeholder training

This training covers the systems and processes used to establish, implement, and maintain the TRM program. The stakeholders in an organization, especially travel, security, and human resources staff, need to know what is expected of them to prevent or respond to an emergency; therefore, this training could apply to any one of them.

Travel professional and TMC involvement in support of TRM programs is covered in greater detail in Chapter 10. TMC training and responsibilities should be limited and should only support the client's TRM program as it pertains to proper, client-provided documentation outlining specifically how the TMC's travel logistics planning, reservations and ticketing processes support the client's TRM program without taking on risk management or mitigation tasks that should be automated and/or performed by the client themselves. Beyond whatever is contained within a client and TMC's service level agreement, the TMC's major focus should be on consistent and reliable data transfer of trip information to tools used by the clients in managing their own program, such as traveler tracking, messaging, pre-trip approval, and other TRM related applications. Employers must always be reminded that their TMC suppliers can be supportive of a limited number of their TRM processes, but the TRM processes are their processes to define, govern, and manage themselves first and foremost.

Crisis management team (CMT) training

This training focuses on simulations and drills to ensure that crisis management plans and procedures are exercised and that people know what is expected of them in an emergency, including the following examples:

- Crisis and response team structure and hierarchy
- Crisis communications and reporting
- Policy and procedure set up and implementation
- Response protocols
  - What can be handled by crisis response call center suppliers under standard protocols without additional approvals?
  - What circumstances require additional approvals, by cost and incident type?
  - What triggers are in place to call the employer's internal crisis response team or select stakeholders into action?
- Continuous process improvement
  - Feedback, measurement/metrics
  - Implementing program changes influenced by feedback, benchmarking and evolving risk factors

Organizationally, employers need to provide a structured and accessible traveler safety training program, that clearly identifies to all personnel via policy and via as many means of communication possible, how to find and access the training. Additionally, employers should clearly identify who is responsible for maintaining and administering the training, and when the training is required or triggered for travelers. In order to maintain such an ongoing training program, utilization of a learning management system (LMS) is ideal, considering that this kind of training needs constant updating because of changing threats around the world, from civil unrest and terrorism, to biohazards and pandemics. These systems also serve as a standardized and easy-to-access method of communication and delivery for such content, versus a manually administered program. Additionally, most LMS programs can track which people have taken which courses and when. The better ones can tell if the people actually took the entire course, versus logging in and not completing it. In the case whereby a traveler is harmed in the course of a business trip, which could have been avoided if skills from proper training had been used, employers will need to be able to prove that not only did they offer mandated training, but whether or not the traveler did or did not take it. If they did not take it, there may be cause for negligence on behalf of the traveler. However, one could argue that the employer should verify that the training was indeed completed prior to travel.

Best practices around when to administer pretrip traveler training, are tied to active monitoring of traveler bookings to high-risk destinations via traveler tracking systems. This is a good example of additional value of TRM systems, well above and beyond simply traveler tracking in the event of a crisis. From a mitigation perspective, regardless of how a TRM stakeholder is notified that a traveler has booked a trip to a high-risk destination, such notification should trigger steps to advise the traveler that upon approval to the destination in question, the traveler must take the company-required training.

Cyber security and social media training safety training

Cyber security and social media safety training should be a part of employer-provided training for all employees, expatriates, and contractors. While it may sometimes seem like overkill to have so much training and to have policies on topics that many people deem matters for common sense, the lack of a position or instruction on certain issues can leave an employer liable. Topics considered should include:

- Prohibited use of any social media using a work-related e-mail address
- Restrictions regarding posting any unauthorized work-related content or information via social media
- Restrictions on sharing work-related travel itineraries via third-party applications or social media for security purposes
- Criteria for traveling with special "clean hardware"
- Authorized and unauthorized use of company-issued resources (mobile phones, computers, etc.) versus personal property (mobile phones, computers, etc.) for business use
- Restrictions regarding installation of unauthorized software on company equipment (phones, laptops, etc.)
- Policies and procedures for protecting company data/intellectual property while traveling or in the office
- Responsible use of e-mail and the Internet
- Best practices to avoid phishing scams and installing or spreading malware

Risk assessment

The purpose of risk assessment is to ensure that each trip or assignment is evaluated or scored for risk, and considered within the overall decision process. However, the concept of risk assessment is not exclusive to the individual trip or meeting. Risk assessment should also be applied to each company's approach to or plan on mitigation, prevention, or reaction to risk in most circumstances.

Risk assessment is the foundation of the overall TRM program and should be conducted on every trip, assignment and special event. For example, Washington, D.C., London, New Orleans, and Madrid may not be considered higher-risk destinations. However, each of these cities has had elevated risk ratings as a result of natural disasters, terrorist events, or civil unrest and protests. While traveling to New Orleans would normally be low risk, if weather forecasters are predicting a hurricane, both the employee and the organization need to be aware of that threat and develop mitigation plans to address it.

Given the changing threat environment around the world, the risk assessment program must take into account both the intrinsic threat level for a destination and any dynamic threats that may elevate the risk of operating in that area over some period of time. This is especially true for employees who are traveling or assigned to a location for an extended time frame.

Organizationally, employers should create a risk management committee. This committee can consist of stakeholders from various departments, depending upon the culture of the organization, but usually includes travel, security, legal, human resources, and facilities. This committee doesn't necessarily set the risk ratings for destinations to which travelers are going, because very few companies have the resources to constantly monitor world events at the analyst level and to write their own security briefings, thus the need for professional, security intelligence and TRM supplier relationships. However, the risk committee sets the tone for the "risk tolerance" across the organization for what they consider high risk, using third-party risk ratings, which in turn sets the tone for crisis response procedures and protocols, policies, and traveler safety requirements for transient travel and meetings, to contractors, expatriates, assets, and supply chain.

As these committees develop their programs, and they mature, this discipline should find its way into developing risk-rating and assessment processes for suppliers and meeting venues, and eventually preferred hotels for transient business travel.

Documenting and communicating each of these activities by the risk committee to key stakeholders is critical, not only to help travelers from a mitigation perspective, but in the event that there is litigation against the company. Having thorough documentation on the steps that the company took to assess the associated risks, along with what they did to mitigate them (and when), will provide the basis for a strong defense against negligence if the efforts are considered "reasonably practical" as compared to their peers in the jurisdiction where the company is being sued.

The following list includes examples of things that a company's risk committee members should take responsibility for relative to assessments:

- Pretrip intelligence and destination-based risk assessments for travelers.
- Insurance such as what insurance covers travelers in high-risk destinations, and what are the provider's requirements for coverage.
- Ongoing area risk assessments for expatriate assignments.
- Ongoing risk assessments for areas where company facilities and projects may be impacted.
- Medical issues such as pretrip vaccinations, known epidemics or pandemics impacting the areas traveled to or where conducting business. Additionally, any special support or services required for preexisting medical conditions, including any laws with regards to traveling internationally with medications that could be restricted, even with a prescription, without proper government approval.
- Physical security such as what access to security facilities and/or protective services does your organization have in the specified destination?
- Supplier assessments (commercial airlines, private charter air carriers, meeting venues, transient preferred hotels, ground transportation suppliers, car rental suppliers, crisis response and support suppliers, medical services and evacuation suppliers, personal and executive security service providers, etc.).

- Plans, conditions, and standards in a crisis to either evacuate or shelter in place.
- Environmental issues such as the likelihood of natural disasters in and around areas to which your company travels or in which it conducts business.
- Conditions and circumstances to serve as the basis for creating and updating all policies, procedures, and protocols relative to TRM.

Risk disclosure

The purpose of risk disclosure is to produce information related to the risk assessment so that all relevant parties are aware of the potential threats that may be encountered.

As an output of the risk assessment process, the organization needs to develop processes to ensure that the appropriate people are aware of the current threat environment. One model is to ensure that employees get basic health and safety information on each trip. If the trip risk assessment level exceeds a defined threshold, then others within the organization should be notified of the potential exposure. Typically, the people to be notified include Travel, Risk/Security, and the line manager.

The risk disclosure process should ensure that each stakeholder understands the nature of the threat, how it may impact the employee and/or organization and what should be done to eliminate or minimize the risk.

Practical application examples of risk disclosures include:

1. Risk report push at time of trip booking—Within a managed program, itinerary data from trips booked via an employer's TMC, can be automatically exported to implemented third-party TRM platforms where the trip destinations are rated against the third-party risk database. The third-party risk platform should automatically generate an e-mail with intelligence-based risk reports on the destinations included within the trip. This process ensures that the "push" of the risk disclosures are automatic, and not subject to human error, and also provides a documentation trail via the e-mail history (when sent and with what content).
2. Applicable alert push—Once a system for transmitting real-time or near real-time PNR (passenger name records) or reservation updates to an employer's TMC is established, it can automatically export that file simultaneously to an employer's TRM supplier. If an alert is issued for a destination included on a traveler's itinerary (both pre- and mid-trip), that alert should automatically be distributed to the traveler via one or more of the following methods:
   a. E-mail (primary)
   b. SMS text message (secondary)
   c. Smart phone application message push (supplemental)
3. Custom messaging (push)—Either a customized or standardized, preset message determined by a rule (e.g., specific destination, supplier, risk rating, etc.) is distributed automatically when the PNR is received by the travel management and travel risk suppliers, or an employer user of a TRM platform solution to one or more travelers listed on a report that was run based upon a particular risk that has been identified by the employer.
4. Intelligence access (pull)—Making an updated, third-party TRM supplier database available for query by travelers on a company intranet. This should never be the exclusive method of disclosure and should be used as supplements to the collective use of practical risk disclosure application examples 1, 2, and 3 above.
5. Training, both general and trip/destination specific.

To a lesser extent, for more general, nontrip–specific disclosures, perhaps relative to policy, procedures, or specific job descriptions, new-hire package paperwork and policies could provide a limited amount of use as risk disclosure mechanisms.

IMPORTANT TO NOTE: Risk disclosures, specifically in examples 1 and 2 above, should always be automated, to eliminate the possibility of human error. All too often, small and midsize companies try to "cut corners" on the cost of TRM by subscribing a few administrative assistants or others to some kind of security alerts, then asking them to monitor the alerts and forward the alerts to those travelers in or bound for locations where the alert may apply. This is a dangerous mistake that could cost companies considerable damages if the alert was wrongly interpreted, or simply forgotten or overlooked.

Risk mitigation

The purpose of risk mitigation is to develop strategies and solutions that will result in a level of risk that is acceptable to all parties (i.e., the employee, the manager, and the company).

Identifying potential threats is not enough. The organization and employee need to understand how relevant a threat is to the trip and business being conducted. From there, both standard and ad hoc mitigation strategies need to be developed to reduce the resulting risk to a level such that both the employee and the organization are comfortable conducting the trip. If an acceptable level of risk cannot be achieved, then alternatives such as canceling or rescheduling the trip can be explored.

Best practices in risk mitigation are more effective if implemented on a pretrip basis, with a plan in place for supporting risk mitigation during the trip, should it become necessary. Only focusing on mitigation after departure and/or post incident is reactive in its approach. Employers must be forward thinking, anticipating risks whenever possible and only outsourcing the mitigation of those risks to qualified support partners, such as TRM and intelligence firms with global networks at their disposal managing and mitigation risks pre and post incident.

Common mistake(s) with risk mitigation strategies include:

- "We don't need a plan or mitigation exercises, because if and when something happens, we will figure it out." In case of litigation, no plan equals no evidence that an employer was prepared to handle the crisis, a strong indication of negligence.
- "Our travelers know which number to call when they need help." That means there are multiple response hotlines for different types of situations and different regions, versus one global crisis response hotline and likely a lack of communication and training on who to call for what.
- Some employers refuse to provide a global crisis response hotline, the kind that supports and provides case management for all company protocols in conjunction with all participating suppliers (medical and security), as well as insurance providers. If someone isn't available in the middle of the night, who immediately knows how the company wants a situation handled, chances are that the crisis won't be handled properly. Additionally, on-call resources that understand how to guarantee payment for services in the event of a crisis is key, because any delay in guaranteeing payment can cause unnecessary damages to the impacted traveler(s). Such delays can be seen as a lack of preparation, and another instance that has the potential to be perceived as negligence.

One best practice for risk mitigation is to incorporate a pretrip approval process, which includes a risk rating for the destinations booked (preticketing), within your managed travel program. Such a process can involve critical stakeholders within the company for enforcing compliance to policies where the predetermined risk threshold or standard for the employer has been set. For those times when a traveler must go to a high-risk destination, having such a process in place should create a documentation that shows that management approved the trip and that steps were taken prior to the trip, based upon the approvals process, to prepare the traveler for minimal risk exposure prior to going, or for what to do in the event that the traveler experiences a crisis.

Risk monitoring

The purpose of risk monitoring is to develop real-time monitoring of world events for potential threats to travelers and expatriates.

Each organization needs to have an around-the-clock (24 × 7) process to monitor the current threat environment across all segments of the trip. This includes travel destinations, modes of transport (e.g., air, rail, sea ports), hotels, transportation carriers, etc. Once a new threat is identified or any existing threat level changes, that new information must be captured and fed into both the risk assessment and notification processes.

Best practice examples of risk monitoring include:

- Daily review of traveler tracking reports or dashboards that include itinerary data, city and country risk ratings, and active alerts commensurate with active trip date ranges.
- Active monitoring of risk exposures relative to PNR trip data and relative risk exposures by a third-party risk management supplier, while incorporating employer policies and protocols.
- As previously mentioned under "Risk Mitigation," an effective pretrip approval process that incorporates destination risk ratings is an effective method of risk monitoring, which should not be used alone, but in concert with other methods.

Risk monitoring—common mistakes

When something happens, our travel agency tells us if we had anyone impacted.

This is a bad practice. TMCs or travel agencies can provide employers with tools and intelligence to support risk management, but they are not risk management companies. Remember that it is the employer's, not the TMC's, responsibility to get ahead of, and respond to any existing or potential risks impacting travelers. Many employers use more than one TMC globally, but whether they use one or several, employers are responsible for ongoing supervision of data quality and the process that feeds critical trip information into TRM solutions. TMCs have many clients, across many countries, and many of those clients do not have globally standardized data. Part of using a TRM technology platform is is to normalize the data into a central monitoring database for the user (tracking travelers from one or more TMCs) to quickly access and run reports to identify if there are any impacted travelers based upon specific events. Sometimes these reports and processes are triggered based upon unique employer policies, or risk criteria, which may not match employer TMC's criteria for crisis communications and running/distribution of incident-based traveler tracking reports.

How should risk monitoring affect your organization? Depending upon what things and circumstances your company monitors, your crisis response and/or business continuity plan should address certain actions triggered by risk monitoring. Some potential triggers could be:

- Active monitoring of reservations data to identify any flight that exceeds the maximum number of employees or executives permitted on any one flight together.
- TRM solutions that monitor incident alerts and compare them to active traveler itineraries. *Example:* Your intelligence provider issues an alert with a severity level of critical, which coincides with identifiable business traveler itineraries. Such a use case could trigger a notification to travelers and key stakeholders to proactively reach out to the affected parties.

Response

The purpose of response is to determine how an organization provides resources and support in the event of a critical incident.

The Response KPA addresses the reactive component of the overall program. An organization needs to be prepared for an event or incident. Having proper response plans and protocols in place with the appropriate resources is the foundation of the Response KPA; however, integrating this into the Training KPA and the Communication KPA are what allow an organization to reach full response maturity.

Types of response protocols could be:

- Steps for facilitating evacuations of travelers, and additional steps for expats.
- Steps for management of medical-related issues while traveling, along with criteria for escalation if an incident's severity goes beyond standard limitations for crisis response support (typically third-party crisis hotlines).
- Steps for managing lost or stolen intellectual property.
- Steps for managing pandemic situations, both on the road for travelers and in corporate facilities.
- Kidnapping and ransom, as well as proof of life, processes.
- Steps for proper handling of a traveler who has been victimized or experienced trauma while traveling for business.

Having access to the resources around the world to support any response is equally important as having plans in place and perhaps a third-party– or even corporate-hosted crisis hotline. Many employers have one or more medical services and evacuation providers that they work with, covering different parts of the world, using potentially multiple insurance providers to pay for such services. However, what many employers overlook is access to security-related resources should they need them. It is a mistake to think that just because you have a global medical services and evacuation program, that you have covered all of the bases with regard to response. While it usually isn't feasible for an employer to maintain its own global network of security-related service providers, such provider networks should be a part of any relationship with a contracted TRM technology and services provider. Reputable TRM suppliers maintain global networks of suppliers who can provide assistance in a variety of areas, from executive protection and security-related evacuations to local support to deal with local authorities and governments, should a traveler be in some sort of legal trouble. Employers shouldn't assume that just because they have medical evacuation coverage that those policies or providers would support security evacuations. Some do, but this is something to be aware of, and to consider when evaluating providers. Additionally, these types of on the ground security networks, in addition to other things, like on-staff security analysts for creation of risk intelligence, are what set apart the true best-in-class TRM suppliers from those companies who want into the TRM market, but only provide technology.

Notification

The purpose of the notification process is to ensure that the appropriate people are informed of any relevant travel risk information before, during, or even after a trip, so they can make rapid and thoughtful risk-related decisions. While another KPA (risk disclosures) is very similar and is focused primarily on disclosing risks to travelers based upon assessments, notification is communication on a broader scale to potentially multiple stakeholders, such as crisis response teams, department managers, C-suite executives, and others for purposes that may trigger processes outlined in a crisis response or business continuity plan. For example, if a kidnapping took place, is there a protocol in place to assemble a specific team of internal company stakeholders, and to follow a plan that has been studied and practiced annually?

An additional example would be if a call came into a company's crisis hotline that required an evacuation that exceeded the cost included in your insurance coverage or standard protocols. Would there be a plan in place to notify the appropriate stakeholders, who are authorized to approve such services and higher-than-expected costs?

Thinking through how people in your organization are notified in a call to action to not just respond or react to an incident, but in advance of it in order to mitigate it, reflects a mature TRM program and an organization that typically has incorporated its TRM program into business continuity planning.

Data management

Data management is the methods by which an organization organizes, protects, and uses sensitive data to keep it confidential and effectively useful in the process of TRM. The data management KPA addresses the overall process of identifying, collecting, storing, accessing, and maintaining this information. This is the KPA responsible for assembling the data that represents a community's prior experience.

The breadth and depth of information collected and maintained to support a comprehensive TRM program can be significant for a program of any size. This information includes personnel contact profiles, trip itineraries, long-term assignments, threat information, destination (country/city) information, organizational structure, etc. A critical part of this KPA is an integrated data quality assurance process that ensures data is accurate and properly archived for use in audits and risk management decisions.

Some examples of key components relative to quality data management in TRM include:

- Human resource (HR) data feeds for use in traveler profile creation and maintenance.
- Standardized traveler profiles globally (i.e., use the same technology for profile management globally, such as an online booking tool or third-party profile management system).
- Globally consolidated travel management program with one TMC (agency), if possible.
- Documented process for capturing all travel itinerary data, whether booked inside or outside of the program, using open booking data capture tools, manual trip entry tools, or direct connect data feeds.

Communication

The purpose of communication is to ensure that all stakeholders understand the TRM program and their role within it.

This KPA focuses on the organization's responsibility to properly communicate the program and all of its elements to each constituent group. These groups include employees, management, senior management, emergency and CMTs, contractors, families of travelers, and external entities such as vendors, suppliers, and channel partners.

These communications cannot consist of one-time distributions. Again, TRM is an ongoing discipline that must be continually improved upon and thus communicated accordingly via different mechanisms such as:

- Training
- Newsletters
- Executive communications

The end result of thorough assessments

While adopting the TRM3 assessment model will take time and resources, the end result provides companies with the following:

- Ongoing documentation or evidence of the company's efforts to make "reasonable best efforts" to meet its duty-of-care obligations, especially if data is benchmarked and showing consistent efforts to improve year over year in a repeatable process.
- A system of measurement, or game plan, for continuously improving safety and security programs and policies that serve as the foundation for mitigating risk and keeping people safe.

Most companies begin the TRM3 assessment process with at least one KPA in the level 1 (reactive) scoring range, plus some other areas that may be more developed. For example, a company without a defined risk training plan, but the other nine KPAs measured as "defined" (level 2) or better, would score a "1 plus 9" score. On average, companies strive to achieve at least a level 3 score or above. Levels 4 and 5 are typically very heavily managed, "lean" or six sigma–type organizations such as government entities, and even then these scores are achieved only by a small percentage of companies that get assessed.

Building a proactive travel risk management program

3

Chapter 2 defines key performance areas (KPAs) and building blocks to be used in assessing your travel risk management (TRM) program or lack thereof. This chapter addresses important questions you should address and provides examples in each of the areas that may be useful when building a TRM program.

The following is a breakdown of TRM program building blocks with important questions and thought-provoking examples in outline form.

Planning

Even though each company is different, and the level of complexity and sophistication for how each company relates organizational goals and policies to TRM varies, a company's planning phase should consider the following:

1. How will your policies affect your program and your suppliers, and their service level agreements?
    a. Will you require your travel management company to commit to a specified process workflow in support of your TRM program?
        i. Risk based, pretrip approvals.
        ii. Identification and "black listing" of suppliers deemed unsafe.
        iii. Processes and timelines for crisis response protocols in the event of a critical incident requiring traveler outreach and reaccommodation.
        iv. Timely pretrip reporting of flights exceeding maximum number of employees allowed on the same flight, as well as special limits on key executives traveling together.
        v. Ensuring that in the event of a crisis, your travel management company knows how to report and refer distressed travelers to the appropriate contacts for any support beyond travel logistics, and to *not* offer any safety and security advice or assistance.
        vi. Verifying that your travel management company is trained on reporting specific risk policy violations identified during the booking process to the company prior to travel (e.g., booking a rental car for use after a redeye domestic flight or overnight international flight).
    b. Will you rely on key airline partners for support when it comes to company needs in the event of a crisis, or is it every man, woman, and child for themselves?
        i. Will they guarantee seats on oversold flights for travelers with qualifying mileage program status or participants in VIP/special services programs? Will their partners comply?
        ii. Which airlines will support you in the event that you require a charter?
            1. How quickly can you arrange one?
            2. Is the process streamlined?
            3. Who is authorized to order a charter, and what is the internal process before placing the order?

  c. Will your annual hotel program incorporate requirements related to safety and security best practices and protocols?

     i. Will your largest preferred supplier hotel chains give you senior-level security contacts that can act globally?

     ii. Do they have designated stakeholders at each property that are both trained and responsible for handling security issues arising from assaults, thefts, and other investigations in conjunction with local authorities?

     iii. In the event of a large-scale critical incident with a shortage of rooms, will suppliers help to "shelter in place" your travelers, even if all rooms are sold out (e.g., use of meeting rooms and other properties for temporary housing during events such as ash clouds)?

     iv. Do the properties adhere to fire safety, evacuation, and appropriate security protocols (e.g., in high-risk destinations, is attention given to how close cars can pull up to the main entrance doors without first passing through security)?

     v. What are their medical emergency protocols, and do they have defibrillators onsite with trained personnel?

     vi. Will you allow usage of shared economy accommodations such as Airbnb, VRBO, and Homeaway?

  d. Ground transportation providers (private cars)

     i. Will they follow security protocols?

       1. Airport or on location "meet-and-greet" services (no individual names on placards; only company names or fake company names).

       2. Provide driver photos, names, and contact info in advance of arrival.

     ii. Do they have sufficient insurance coverage?

     iii. Will you allow shared economy-type providers like Uber and Lyft?

  e. Corporate car rental providers

     i. Will they provide contracted insurance coverage for best available rates?

     ii. What is your contingency plan for those locations where contracted insurance coverage is excluded from the terms of your agreement?

  f. TRM Suppliers

     i. Do they provide a single number for all travelers to contact for any kind of emergency (medical, security, theft, translations, trouble with local authorities, etc.)

     ii. Have you established detailed security protocols and contacts for how to handle any situation, including, but not limited to:

       1. Key global and regional contacts for different types of crises, and when to contact or escalate decision support.

       2. Under what circumstances and amounts can the provider handle a situation and authorize payment without authorization and/or escalation?

       3. Do they have all of your "Business Travel Accident" or other insurance policies and contact info for coordination and case management under appropriate circumstances, or can they provide any of the following (emergency medical service providers, security-based service providers, kidnap and ransom service/insurance providers)?

       4. Do they have specific use cases outlined whereby a process would be triggered that would involve traveler outreach for safety purposes (e.g., safety outreach and communications with travelers identified as directly impacted by an event such as natural disaster)?

       5. Are there separate protocols and procedures in place for expatriates and family members, versus transient travelers?

2. Response planning and preparation
   a. Documented plans for how to handle the following:
      i. Pandemics or biohazards
      ii. Physical assaults
      iii. Transportation accidents (airline crashes, car rental accidents or rail derailments)
      iv. Civil unrest
      v. Arrest by local authorities
      vi. Kidnapping
      vii. Natural disasters
      viii. Incidents impacting business continuity

Training

There are several aspects and levels of training related to TRM that companies must actively develop, administer, and manage, including the following:

1. New-hire policy and duty of loyalty training
   a. New-hire risk policy orientation
      i. Thorough travel and security policy and process review and documentation in all new-hire packets
      ii. Business Travel Accident and any other relevant insurance disclosures
      iii. Safe traveler training prior to conducting business travel, including crisis response program information
         1. Online and offline authorized booking process
         2. Pretrip approval process for high-risk destinations and/or other established criteria
         3. Hotel fire safety training
         4. Ground transportation safety
         5. Female business traveler training
         6. Crisis response training
            a. Who to call
            b. What to do
            c. When to do it
      iv. Liability waiver for travel and/or security policy violations (e.g., open booking)
      v. Signed acceptance of all training and policies
   b. Duty of loyalty training
      i. Clearly documented responsibilities for travelers, contractors, expats, and accompanying family members for:
         1. Ethical behavior
         2. Cultural sensitivity
         3. Policy compliance
         4. Compliance with legislation and law enforcement in applicable jurisdictions
   c. Destination-based training
      i. Specialized training for travelers prior to going to places like parts of the Middle East, Africa, and even China, when available
      ii. Biohazard awareness (e.g., Ebola)

      d. Electronics and intellectual property training
         i. Safety precautions for devices such as:
            1. Laptops
            2. Smart phones
            3. Tablets
         ii. Internet connections
         iii. File sharing and access to sensitive documents via:
            1. Cloud computing
            2. Flash drives
            3. Device hard drives
            4. E-mail
            5. Virtual private networks
      e. Survival training
         i. During and after an abduction or kidnapping
         ii. Natural disaster
         iii. Assault and battery
         iv. Self-defense (when necessary)
         v. Active shooter
2. Travel policy, process, audit, and safety training for corporate administrators
      a. Administering and supporting policies
      b. Identifying policy noncompliance and violations before, during, and after trips (expense reports), and the traveler's responsibility to disclose
      c. Support of pretrip safety precautions as part of standard travel planning checklist
      d. Standardized small- and large-meeting processes and safety protocols with authorized meeting and event planning suppliers who contribute and support the company's TRM program
3. Travel consultant training (from travel agency or travel management company [TMC])
      a. Certification process for agents on company's travel policies regarding cost, preferred supplier, process, and safety
      b. Clearly outlined and understood processes for bookings that either do or do not require any exception approvals
         i. Trigger reporting and/or approvals process when necessary, as dictated by policy or service-level agreements
      c. Understanding where to direct distressed travelers in the event that they receive calls from someone needing more than travel logistics/reservations assistance, including noncommercial travel evacuations assistance
      d. Crisis response process in the event of a "critical incident" (usually defined as an event that has the potential for severe bodily injury, death, or mass travel delay), which may include:
         i. Global reporting and distribution on potentially impacted travelers to key stakeholders
         ii. Specialized communications and reporting to key regional stakeholders, under specified circumstances
         iii. Communication of "all clear" when no potentially impacted travelers are identified
         iv. What to do in the event that there are potentially impacted travelers, including, but not limited to, reservations assistance, special approvals (e.g., for business or first class), etc.
4. 24 × 7 Monitoring
      a. Self-service monitoring using:
         i. TMC traveler tracking solutions
         ii. TRM company traveler tracking solutions (all requiring TMC data)

b. Full service or automated monitoring using:
   i. Rules-based automation to automatically generate and distribute reports containing traveler itineraries of travelers who are considered at risk using third-party risk-rating content
   ii. Crisis management staff from contracted TRM firms, actively monitoring traveler movement before and during trips, to identify any potential need for traveler outreach and support

First-generation traveler tracking software were merely "dots on a map" based on travel agency/TMC reservations data alone. A decentralized travel program would find benefits from using a TRM firm's platform, which can consolidate, scrub, and scrape nonstandardized reservations formatting from multiple travel providers into one database. Although this still works well with a consolidated program, often best practice dictates that the most effective travel programs are consolidated with one global travel management company, in which case the TMC's solutions can be considered, sometimes at a more competitive price point. However, buyers must beware when purchasing any traveler tracking solution from a non-TRM/consultancy firm, unless the solution in question is deeply integrated with intelligence and crisis response networks and services from a reputable TRM leader. Because TRM is such a hot topic of discussion, and companies now realize that they must have something in place, there are many solutions providers in the market bidding for a piece of the potential market. The majority of solutions not provided by or not in deep collaboration with one of a few world-class leaders in TRM are insufficient to truly support a comprehensive duty-of-care/TRM program. They are sold more as products than as solutions, and TRM programs must always be designed and maintained as solutions-based programs, not merely software-based.

More recent versions of traveler tracking/monitoring solutions incorporate either GPS (latitude/longitude) locations or corporate credit/charge card preauthorization data, as a supplement to TMC reservations data. A traveler could land in London on a Tuesday, but be 200 miles away on Thursday before the traveler's return flight home on Sunday. These additional data sources can be a huge help in locating travelers more precisely in an emergency.

However, it's not enough to simply know where your travelers are. The ability to consolidate or match travelers' location data with active alert intelligence (issued by true intelligence analysts, not newswires or networks) is critical for efficient monitoring and proactive traveler risk management. Of course, one can receive an alert or see something on the news and run a report, but having the ability to automatically run a report anytime an alert is issued that matches your travelers' locations, is immensely more efficient. Additionally, these systems commonly allow the users to communicate via e-mail and/or SMS text message with travelers listed on reports, and can also actively forward alerts or destination specific risk reports relative to their travel itineraries, which are "must haves" in terms of a company's "duty-to-disclose" responsibilities, respective to their duty of care.

The distribution of these disclosures (alerts and risk intelligence reports) must never be done manually because manual disclosure would be subject to human error. This cannot be stressed enough. Do not attempt to cut corners and save money by subscribing a travel arranger, administrator, or anyone else to an alert subscription and depend upon them to manually review and distribute these to the appropriate travelers. This is a very bad practice, and subjects the company to needless liability. These kinds of disclosures should be sent *always* automatically via e-mail, at the time of booking travel reservations or at the time changes to a booking are made that introduce new destinations to the itinerary plans.

The following is likely repeated several times within this text, but is worth repeating because it is so common. It is commonly believed by companies that they adequately manage risk by knowing where their travelers are (traveler tracking), and by administrative assistants manually forwarding alerts to travelers via e-mail, or even asking their travel agents to do it. This approach is not only a bad idea, it is an incomplete solution, failing to address each of the building blocks of a proactive TRM program, as discussed in this chapter. Travel agents, who get involved with manually forwarding these types of information on behalf of their customers, are not only taking on some of their client's duty-of-care responsibilities, but are also assuming major liabilities as well, establishing themselves as the source for the delivery, timeliness of delivery, failure to deliver, and the accuracy of any information shared, which is a 24/7/365 days a year responsibility.

Crisis and incident response hotlines

Historically, some companies used to view a medical emergency hotline as sufficient in terms of what they provided to their travelers in the event of an emergency. However, we now know that the definition of emergency has greatly been expanded and the full use of a best-in-class crisis response hotline can even extend beyond TRM, and can incorporate aspects of operational risk management (ORM) for things such as facilities and supply chains. However, from a TRM perspective, travelers need one number globally for any kind of emergency, including, but not limited, to the following:

1. Medical emergencies and evacuations
2. Security emergencies and evacuations (civil unrest, natural disasters, harassment, etc.)
3. Emergency translation services
4. Assistance with local authorities (police, local and federal governments, consulates and embassies)
5. Executive protection
6. Secure ground transport coordination
7. Kidnap and ransom incidents

While these hotlines may provide case management and some of these services through their own provider networks, they should also work with contracted service providers and insurers on behalf of the company, as part of their customized security protocols as defined and documented during the implementations process of a crisis hotline. Custom protocols could include, but not be limited to:

1. Collaboration, case management, and billing assistance with third-party insurers and service providers, such as medical and K&R (kidnap and ransom) suppliers
2. Preauthorization of services up to predefined spending limits, based upon company-defined criteria
3. Who to contact when situations exceed preauthorization limits
4. When to wake up key security contacts in the middle of the night for decision support
5. Based upon location
6. Based upon type of incident
7. Based upon cost

A common question for this type of consolidated hotline is "Why do we want a 'middle man,' when we have some of these relationships in place already?" The following items are just a few answers that address this type of question and explain why such a number is beneficial.

- One number for all employees, contractors, and expats is more efficient and less confusing than several different numbers for different requirements.
- All calls should be recorded and managed as cases with detailed reporting on all conversations, requests, service orders, and activity related to the incident. This is typically exactly what employer's need when in the midst of litigation.
- Facilitates preauthorized support services when needed.
- Provides vetted supplier networks for services required, when the client doesn't have a provider or insurer in place for the incident in question.
- Billing/invoicing assistance when necessary.
- Customized protocols (processes) based upon your company's culture, needs, and risk thresholds.
- Case management should mean that the hotline provider doesn't just transfer calls to the appropriate customer contracted supplier, but sticks with the traveler through the entire process from start to finish, until the traveler is safely home, documenting each party's involvement in managing the incident. Such documentation is necessary in cases involving litigation.

Feedback

Establish information collection points, whereby you can document and review similar incidents over time and create your own continual process improvement workflow. Information collection points include:

1. After a crisis communication distribution:
   a. Natural disasters
   b. Civil unrest
   c. Biohazards
   d. Terrorist activity
   e. Transportation disputes/disruptions
2. Posttrip, traveler interviews after returning from high-risk destinations.
3. Incident rehearsals or crisis response exercises with key suppliers.

Whether you follow Six Sigma or other methodologies, you must have a defined process, controlled variances, and, most of all, data (feedback) to measure and analyze.

When it comes to creating a well-rounded and proactive TRM program, further to the Six Sigma reference above, to have continual process improvement with consistent feedback and data, companies must approach TRM as a holistic and highly managed solution that is critical to their ongoing success, reputation, and operational readiness. One cannot simply buy some technology and believe that a "turnkey" approach is sufficient. It is not. The best programs that most effectively manage duty of care, are a collaborative collection of suppliers, working together under a managed program structure with one or more key stakeholders within the organization with responsibility for TRM. This will require a noticeable investment, but consider the cost of not being prepared and being found negligent. If you were a shoe manufacturer, you would have to sell an awful lot of sneakers to cover the likely damages paid to a claimant in the event that the claimant prevails in a duty-of-care case against you.

Common mistakes companies make when starting out with their first TRM program include:

1. Travel Insurance—Using consumer-based travel insurance products for broad, enterprise-wide use. Not everyone who inherits the responsibility of managing travel within an organization has travel industry experience, and sometimes when faced with decisions early in their training about the kinds of products to source, they refer to consumer-based products, such as trip-based travel insurance. These kinds of products often have coverage not needed by a company (may be covered in other policies, if at all), charge excess premiums based upon age or preexisting conditions, and many will not preauthorize or arrange direct payment to hospitals and clinics abroad, requiring the customer to pay for all expenses in advance and file for reimbursement upon returning home. This could result in many thousands of dollars in out-of-pocket expenses without any guarantee of payment for those expenses deemed ineligible. Additionally, if a delay in payment results in delayed or denied treatment, employers could end up liable for additional damages.

2. Services, Support, and Gaps in Coverage—Emergency medical memberships or "network access" types of programs can be sold at a company level as well as at an individual level, and are often mistaken as insurance. While membership sometimes covers the cost of medical evacuation, and access to medical advice or referrals, they sometimes do not cover medical costs. Sometimes these kinds of memberships are purchased with this in mind because a separate insurance plan indemnifies the company from any expenses incurred through the medical membership's referral/supplier network. Other times, clients can opt to "self-insure" and pay for expenses directly as incurred, or buy a supplemental policy with the membership to cover expenses. The mistake is not understanding the differences between insurance and network access, along with gaps in coverage.

3. Monitoring Only International or High-Risk Trips—Because so many TRM-based traveler tracking and disclosure products charge using a per-booking fee structure, companies get stingy and think that they can save a few bucks by negotiating a deal where they only monitor and pay for international trips or trips to high-risk destinations. This is wrong, for many reasons, such as:
   a. A crisis can happen anywhere, from natural disasters and civil unrest, to acts of terrorism.
   b. Major events can and do happen in low- to moderate-risk locations quite often.
   c. If you are only actively monitoring international travelers, how can you effectively protect and fulfill your duty-of-care obligations to those domestic travelers, in from out of town, in the same city as your international travelers, where a bomb may have just gone off?

4. Insufficient Crisis Support—Having more than one number for travelers to call in a crisis (e.g., one for medical issues, another for security issues, etc.), or not having sufficient crisis support because you think that a medical services hotline alone is sufficient.

5. Overlapping or Extra Insurance—Understanding what a company already has in place in the form of business travel accident or other forms of corporate insurance, and working closely with your legal, human resources, and finance departments, can help to avoid the possibility of being overinsured.

6. "This Won't Happen to Us!" Syndrome—Putting off or avoiding the implementation of a proof-of-life policy and kidnap/ransom insurance coverage.

Travel risk policies, compliance, and supplier safety

4

Travel policies were originally designed as "travel expense" policies, that is, the rules relative to reimbursement for approved business travel expenses. Unfortunately, today, even with decades of formal corporate travel management associations and best practices in existence, a large percentage of companies still primarily focus on procurement or cost-containment measures, but often overlook risk in their policies.

While travel policies may largely consist of purchasing behavior, expense limits, preferred suppliers, and rules for reimbursement, today they must both weave safety and security-related content into the travel policies whenever possible, and potentially devote an entire section to risk management and traveler safety.

Before you begin drafting a new travel policy, or begin adding important risk components to an existing policy, you must first have the following in place:

1. C-level support for the company's duty-of-care initiatives, including:
   a. Communications to the entire company about the importance of the company's risk management initiatives, explaining how the policies and procedures apply to everyone and how the C-level sponsor will personally support compliance enforcement, but that the named administrative stakeholders are fully authorized by the C-level sponsor to enforce policies on the company's behalf. There may be alternative policies for different kinds of travelers, but overall, there are risk-related policies and procedures that apply to everyone. This level of personal endorsement and specific empowerment of individuals to manage policy compliance is the root cause of any policy's compliance success.
   b. Assignment of key stakeholders to a "duty-of-care" taskforce from each of the following departments (if applicable):
      i. Risk/security department—If you have a risk/security department, these are the stakeholders who will likely take the lead responsibility for active monitoring of travelers and travel-related risks based upon intelligence provided by your risk management supplier. These are also the stakeholders who will set standards for pretrip preparation and training, as well as crisis response protocols. Unfortunately, most small and midsize companies do not have dedicated risk employees or officers, so without a risk department as the project lead, the responsibility often falls to the travel department, in close concert with other departments.
      ii. Travel department—Because the foundation of traveler tracking and proactive risk monitoring is travel reservations data, the travel department is uniquely positioned to play a hands-on role in implementing, administering, and managing risk-related policies and procedures in some capacity. The department's ability to manage and standardize the reservations data for use in risk management systems is paramount to getting a travel risk management (TRM) program off of the ground. The best way to manage risks is to prevent them altogether. One can attempt to do this with an operational process in place, supported by your travel management company (TMC), for a pretrip approval process that incorporates risk ratings for the

destinations where your travelers plan on visiting, before they purchase the tickets. Travel departments are usually the ones that drive, administer, and document travel policy compliance and violations. Corporate travel manager job descriptions initially were all about traveler customer service, travel department operations, preferred supplier, and contract sourcing and utilization, plus cost containment. Subsequently, the descriptions expanded to include such things as corporate governance compliance (i.e., Sarbanes-Oxley Act of 2002 [SOX] compliance for U.S. public companies) and expense management (corporate card administration and expense reporting). Today, these roles have begun to take on risk management responsibilities in concert with other departmental stakeholders, and are sometimes the lead stakeholder in the absence of a risk/security department.

iii. Finance—One of the best drivers of compliance is often controlled by the finance department. This is the ability to refuse reimbursement for expenses incurred in violation of policy. Specifically, the most effective driver as related to TRM is to refuse payment of expenses booked outside of the authorized booking channels (contracted TMCs or travel agencies). As addressed in Chapter 9, booking via nonauthorized travel providers or websites (open booking) can undermine the company's ability to keep travelers safe by knowing where they are at any given time in relation to their itineraries. Data capture technology for open bookings are ineffective when it comes to ensuring the capture of 100 percent of reservations data without being subject to human error.

iv. Human resources—The documentation of each traveler's acknowledgement of policies and any policy violations in the traveler's personnel file, can be critically important for a company's defense during litigation. Such records can establish acceptance of policies as terms of employment and also establish a pattern of violations, which may or may not have contributed to unnecessary risk.

v. Legal—In addition to making sure that the company policies are appropriately defined and managed in concert with the laws and regulations of the applicable government jurisdictions, the legal department plays a key role in defining the company's "risk tolerance" or culture toward risk, which drives how strict or lenient risk policies tend to be. For example, at what risk rating levels are risk report disclosures and pretrip training required?

vi. Communications, marketing, and training—Consistent internal messaging regarding safety-related policies and best practices, as well as training, is something that keeps the topic at the forefront of everyone's mind, and usually takes some planning and forethought. Utilizing standardized or customized online training videos or courses is helpful, but some companies are entertaining the "gamification" of safety training in the form of video game–based training modules with scorecards and incentives.

vii. Sales and/or professional services—Because these departments typically have more travelers than others, having a senior stakeholder from sales and professional services as part of the team setting and enforcing policy generally promotes compliance and buy-in.

viii. Facilities—More for operational risk management and business continuity planning, facilities departments can ensure that all permanent and temporary company locations meet fire safety standards; have published evacuations and shelter in place training and plans; have access to local emergency fire, police, and medical support services; and that when selecting locations, local crime rates and access to transportation and accommodations are considered.

2. Understanding of cultural conditions and business factors
    a. Is your company doing business in high-risk destinations?
    b. Are you doing business with any form of government (additional compliance requirements)?
    c. Do you have a high number of expatriates living abroad?
    d. Do you conduct a large number of offsite meetings and conferences of all sizes?
    e. Do you have access to documentation for any and all insurance coverage related to business travel (business travel accident insurance, emergency medical and evacuation insurance, security services and evacuation insurance, kidnap and ransom insurance, etc.)?
    f. Do you have a history of a high number of medical claims or lawsuits relative to duty of care? If so, how many per year? Do you have documentation? (This will help you obtain insurance coverage if you do not already have it.)
    g. Has your company's reputation suffered as a result of damages awarded a claimant or for cases of neglect? (These might be reasons for extra precautions.)
    h. Have you conducted a thorough, benchmarked assessment of your existing approach to TRM, identifying areas for improvement?

Sample best practices to support policy development

General travel and reservations

1. Controlled/limited open booking—You cannot help travelers if you don't know where they are. For any exceptions, such as customer-paid travel, a formal data handoff or manual data entry process must be adhered to. No matter how many suppliers (airlines, online travel agencies, etc.) establish direct connections to traveler tracking systems in support of open booking, companies that do this will never be able to ensure that all suppliers globally are supported. Without these direct connections, traveler tracking systems are dependent upon travelers to e-mail itineraries booked outside of their authorized TMC to a program that parses the itinerary into the tracking database. Such actions are subject to human error and should not be entertained, versus mandatory usage of the company's contracted TMCs.
2. No personal extensions or side trips from business trips (employer may be held liable).
3. Code of conduct application—Conduct must defined for a trip from the moment the traveler leaves home on business, until the traveler arrives back home at the end of the traveler's trip.
4. No spouse or companion travel on business trips without express written authorization, which should include a signed indemnification from traveler to company for spouse/companion liability against the company.
5. Make photocopies of passports, leaving copies with emergency contacts and human resources, and taking one with you on international travel.
6. Carry extra medicine on business trips in excess of 7 additional days if possible. Always check whether the countries to be visited have restrictions for certain types of illnesses or medicines entering their borders, and if special preauthorization prior to arrival is required.

Risk disclosures and crisis response

1. Third-party, nongovernmentally sponsored safety and risk intelligence should be made available to all travelers, and travelers must be told how and when this information is available within the policy, such as:
    a. Link to third-party database via company intranet
    b. Links to third-party database via traveler itineraries

    c. Push of alerts and risk intelligence via e-mail based upon itinerary destinations (*always* provide risk disclosures to travelers via *both* push and pull communications, documenting provision and access to information within policy)

2. Document emergency medical services and insurance provider details within policy, outlining coverage and exclusions where possible

3. Provide a single phone number for all crisis response support to travelers for all medical and nonmedical emergencies, including, but not limited to:

    a. Medical emergencies and evacuations

    b. Personal safety emergencies and evacuations

    c. Loss of property or intellectual property

Foreign corrupt practices act

According to the United States Department of Justice (http://www.justice.gov), the following describes the Foreign Corrupt Practices Act (FCPA), which must be addressed in all travel policies after discussions with counsel as to the best way to address in a policy format.

Foreign corrupt practices act—an overview

The Foreign Corrupt Practices Act of 1977, as amended, 15 U.S.C. §§ 78dd-1, et seq., was enacted for the purpose of making it unlawful for certain classes of persons and entities to make payments to foreign government officials to assist in obtaining or retaining business. Specifically, the antibribery provisions of the FCPA prohibit the willful use of the mails or any means of instrumentality of interstate commerce corruptly in furtherance of any offer, payment, promise to pay, or authorization of the payment of money or anything of value to any person, while knowing that all or a portion of such money or thing of value will be offered, given or promised, directly or indirectly, to a foreign official to influence the foreign official in his or her official capacity, induce the foreign official to do or omit to do an act in violation of the official's lawful duty, or to secure any improper advantage in order to assist in obtaining or retaining business for or with, or directing business to, any person.

Since 1977, the antibribery provisions of the FCPA have applied to all U.S. persons and certain foreign issuers of securities. With the enactment of certain amendments in 1998, the antibribery provisions of the FCPA now also apply to foreign firms and persons who cause, directly or through agents, an act in furtherance of such a corrupt payment to take place within the territory of the United States.

The FCPA also requires companies whose securities are listed in the United States to meet its accounting provisions. See 15 U.S.C. § 78m. These accounting provisions, which were designed to operate in tandem with the antibribery provisions of the FCPA, require corporations covered by the provisions to (a) make and keep books and records that accurately and fairly reflect the transactions of the corporation and (b) devise and maintain an adequate system of internal accounting controls.

Penalties for FCPA violations can be steep, including up to US$2 million per violation of the antibribery provision and $25 million per accounting provision violation.

Additionally, civil actions can be taken against the firm or individuals within the firm, including personal fines up to US$100,000 and prison terms up to 5 years for officers, stockholders, directors, or employees who willfully violate the antibribery provision. Willful violators of the accounting provision could face up to 20 years in prison and personal fines of up to US$5 million.

The question could be asked, "What does FCPA have to do with TRM?" Because business travel often involves transborder transportation, the business at hand is being done at an international level, requiring caution and awareness to avoid even the potential for an FCPA violation. With the Dodd–Frank Wall Street Reform and Consumer Protection Act legislation, incentives are provided to whistle blowers. In addition to potential fines, the costs to companies can extend to damaged reputations as a result of media coverage and significant loss of stockholder share value. Consider anything that is being paid for a nonemployee's benefit, that is usable by a nonemployee, or that benefits a nonemployee who could influence business to the benefit of your company as having the potential to violate the FCPA.

Example

U.S. Securities and Exchange Commission charged BHP Billiton with violating FCPA at Olympic Games

According to the U.S. Securities and Exchange Commission (SEC), BHP Billiton was charged with violating the FCPA when it sponsored attendance of foreign government officials at the Summer Olympics. An investigation by the SEC discovered that BHP Billiton failed to create and maintain sufficient internal controls over its global hospitality program connected to its sponsorship of the 2008 Summer Olympic Games in Beijing, China. BHP Billiton had invited 176 government officials and employees of state-owned organizations to attend the Olympic Games at the company's expense, which paid for 60 guests along with some spouses and others in attendance. Many of the guests were from countries in Asia and Africa, who received 3- and 4-day hospitality packages with event tickets, hotel accommodations, and sightseeing services valued at US$12,000 to US$16,000 per package.

"BHP Billiton footed the bill for foreign government officials to attend the Olympics while they were in a position to help the company with its business or regulatory endeavors," said Andrew Ceresney, Director of the SEC's Division of Enforcement. "BHP Billiton recognized that inviting government officials to the Olympics created a heightened risk of violating anticorruption laws, yet the company failed to implement sufficient internal controls to address that heightened risk."[1]

---

[1] U.S. Securities and Exchange Commission, "SEC Charges BHP Billiton With Violating FCPA at Olympic Games" (press release), May 20, 2015, http://www.sec.gov/news/pressrelease/2015-93.html.

Based upon the SEC's order instituting a settled administrative proceeding, BHP Billiton required business managers to complete a hospitality application form for any individuals they intended to invite to the Olympics, including government officials. BHP Billiton didn't clearly communicate to employees that no one outside of the business unit submitting the application would review and approve each invitation. The company failed to provide employees with any training on how to complete the forms or evaluate bribery risks of the invitations. Because of these and other failures, many of the hospitality applications were incomplete or inaccurate, and BHP Billiton extended Olympic invitations to government officials connected to pending contract negotiations or regulatory business, such as the company's efforts to obtain access rights.

Duty of loyalty

While employers have a "duty of care" to employees and/or contractors traveling on their behalf, the concept of responsibilities or obligations of the travelers to comply with safety precautions or policies, putting the company's interests ahead of their own personal interests, is known as "duty of loyalty."

According to Wikipedia, the definition of "Duty of Loyalty" is as follows:

Duty of Loyalty is a term used in corporation law to describe a fiduciaries'
"conflicts of interest and requires fiduciaries to put the corporation's interests
ahead of their own. Corporate fiduciaries breach their duty of loyalty when they
divert corporate assets, opportunities, or information for personal gain.

It is generally acceptable if a director makes a decision for the corporation
that profits both him and the corporation. The duty of loyalty is breached when the
director puts his or her interest in front of that of the corporation.[2]

In addition, duty of loyalty (in context with TRM) means employees should exercise good judgment in keeping themselves safe while traveling for business by not taking unnecessary risks driven by personal interests, such as using a nonpreferred supplier simply because of frequent flyer benefits. Duty of loyalty can also be reflected by following company policies and procedures outlining best practices for safe and security travel, utilizing programs and resources made available to employees, such as vetted service providers and crisis response/support hotlines, and employing safety precautions taught in training classes that contribute to an increased level of precaution and safety. Taking unnecessary risks or purposefully not following safety policies and procedures can breach that duty of loyalty, but it isn't always so evident or obvious unless an employer outlines exactly what that traveler's "duty of loyalty" is via documentation. This is critically important when outlining individual

[2] Wikipedia, "Duty of loyalty," August 26, 2015, http://en.wikipedia.org/wiki/Duty_of_loyalty.

responsibility in your TRM program. Even with very clearly defined duty-of-loyalty responsibilities, not everything can be documented as such. Consequently, duty-of-loyalty responsibilities are often implied by virtue of compliance with the TRM program itself. Such compliance, or agreement to comply and meet those duties of loyalty obligations, can be well-served when employers require some written form of acknowledgement from travelers as to their roles and responsibilities. Some examples of these duty-of-loyalty responsibilities, as a reflection of a documented process, policies, or program components include, but are not limited to, the following:

1. Booking all travel arrangements via approved TMCs, in association with the company's TRM program. If employers do not control the booking data, they cannot easily track and mitigate risks associated with business travel.
2. Using only preferred car rental suppliers and following outlined procedures to confirm that corporate insurance coverage is active and applicable.
   a. Understanding how to obtain required insurance coverage with nonpreferred car rental suppliers in the event that preferred suppliers are not available.
3. Using preferred hotel properties deemed acceptably safe by company risk management stakeholders.
4. Taking required safety training prior to travel destinations that require such training per policy.
   a. Employing safety precautions taught via pretrip security training courses.
5. Following safety and security protocols for ground transportation in high-risk destinations (contracted services that do not use company or individual name placards, and that provide advance information on drivers and vehicles).
6. Reporting any supplier conduct that does not adhere to safety and security requirements of the employer.

Employers could document any number of specific items or implied responsibilities for employees, depending upon how far they want to take the issue, and the extent their legal and human resource departments will support such steps as reasonable.

Supplier safety

Air travel

When a consumer or business traveler checks availability and pricing for airline tickets to an international destination, how much thought or consideration generally goes into whether or not an unfamiliar airline that may offer the best price is safe? Even when a traveler questions the safety of a particular airline, it is the employer's responsibility to make a determination of safety, and to allow approved business travel to be purchased and take place.

How can employers make informed decisions about which commercial airlines are safe and which ones are not? While there is no single, global authority that governs all commercial airline safety, several resources are available to assist in making these evaluations. Bear in mind that employers can greatly benefit in saving time and resources by purchasing intelligence-based airline safety reports from providers who monitor the following resources to produce safety recommendations; however, it is useful to understand best practices with regards to how airline safety is measured and where the best sources of information on the subject can be found.

Recommendations for evaluating airline safety

Step 1:   Evaluating Airline Safety—Country-Level Airline Regulation:

CAAs (civil aviation authorities (CAAs)—A country's governing body responsible for regulating and overseeing that country's airlines. Effective CAAs are responsible for setting appropriate safety and maintenance standards, enforcing compliance with standards, and the ability to shut down operations if compliance standards are not met. (The United States Federal Aviation Administration [FAA] is an example of a CAA.)

While an airline's safety rating could be influenced by a CAA, much depends on the credibility and rating of the CAA, based upon CAA ratings and the CAA's ability to set, maintain, and enforce standards. Some CAAs may fall short in their standards for their respective countries, but may have airlines that meet or exceed industry standards and are recognized as safe, while being under the supervision of a substandard CAA.

Step 2:   Evaluating Airline Safety—CAA Assessments

Some country's CAAs don't have the resources or standards to properly support airline safety and maintenance standards. Because of this, CAAs should be subject to audits or standards that deem whether or not they are trustworthy. The following resources can be useful in evaluating CAAs.

United States—FAA's International Aviation Safety Assessment (IASA) program. The IASA program rates an individual country's CAA on a pass (category 1) or fail (category 2) system, and is updated on an ongoing basis, measuring compliance with the safety standards established by the ICAO (International Civil Aviation Organization). However, this program is specific only to carriers that operate or seek to operate in the United States, or code share with U.S. carriers. Information on IASA can be found at http://www.faa.gov/about/initiatives/iasa/.

Europe—List of airlines banned within the European Union (EU) (sometimes referred to as the EU Airline Blacklist). Operated by the European Commission, this list is updated continuously and is best used for the evaluation of CAAs versus individual airlines. However, this list does provide exceptions for airlines that have met EU standards, even though their CAA did not. This document does have regulatory influence for the EU, and can be found at http://ec.europa.eu/transport/modes/air/safety/air-ban/index_en.htm.

United Nations—ICAO Universal Safety Oversight Audit Program (USOAP). The ICAO is a United Nations (UN) specialized agency that was created upon the signing of the Chicago Convention (Convention on International Civil Aviation). The ICAO works with 191 Convention member states and global aviation organizations to develop international Standards and Recommended Practices (SARPs), which States reference when developing their legally enforceable national civil aviation regulations. The complete list of member states can be found at http://www.icao.int/about-icao/Pages/member-states.aspx. As of October 2013, 96 percent of all Member States having oversight responsibility for 99 percent of all international air traffic have completed the ICAO's comprehensive systems approach (CSA) audit. In addition to these audits, the ICAO has adopted a "continuous monitoring approach" (CMA) to allow the collection of more regular information regarding the level of safety oversight provided by ICAO member states in addition to supplemental, regularly scheduled audits and information collected by relevant external stakeholders and other CMA activities, which as an ICAO Coordinated Validation Mission (ICVM)ICAO Coordinated Validation Mission (ICVM) is performed to validate whether any previously identified safety issues have been resolved appropriately or mitigated, potentially originally identified by an audit.

The USOAP audits focus on a member states ability to successfully and consistently implement the critical elements of a safety oversight system and the degree to which the member state has implemented the ICAO's safety-related SARPs and relative guidelines, procedures, and documentation. The program monitors the following eight components of a member state's aviation system[3]:

1. Primary aviation legislation and associated civil aviation regulations
2. Civil aviation organizational structure
3. Personnel licensing activities
4. Aircraft operations
5. Airworthiness of civil aircraft
6. Aerodromes
7. Air navigation services
8. Accident and serious incident investigations

Information on USOAP can be found at http://www.icao.int/safety/Pages/USOAP-Results.aspx.

The ICAO publishes an annual safety report, and the 2015 issue can be found at http://www.icao.int/safety/Documents/ICAO_Safety_Report_2015_Web.pdf.

Global—iJET International's Airline Safety Newsletter and Worldcue Airline Monitor ongoing, intelligence-driven reports on hundreds of airlines and their associated safety ratings as "Preferred" or "Non-Preferred," as rated by iJET's security analysts, using a host of resources in addition to the ones previously listed in this chapter.

Step 3: Implement policies and program that utilize intelligence and best practices, including, but not limited to the following:

1. Obtain regularly updated airline safety intelligence (minimally on an annual basis, if not more frequently updated).
2. Provide list of airlines not meeting company safety requirements to TMC for monitoring. Use of said airlines should be prohibited.
3. Consider a pretrip approval process for all travel or international travel, and incorporate an approved airline safety validation process (via online booking tools, quality control technology, or as a last resort, manually).
4. No more than five to six travelers on the same flight (C-level executives should never fly together). This number varies from company to company, and many companies have multiple versions of this policy based upon title and/or position.
5. Use of private or chartered air transportation is prohibited without written approval and submission of the following criteria:
   a. Whom aircraft was hired or chartered from.
   b. Type of aircraft (no single engine aircraft allowed).
   c. Proof of provider's liability coverage and amounts.
   d. Proof of valid pilot's license and credentials.
   e. Disclosures of provider's and pilot's safety/incident record.
   f. Departure and destination points, with departure dates and times.
   g. Number of employees and total travelers.
   h. Detailed passenger manifest.
6. Training
   a. Exercises and movement to avoid deep vein thrombosis.
   b. Sanitation/cleanliness—lavatory and seating areas (recommend bleach-based sanitizing wipes).

[3] International Civil Aviation Organization, "F.A.Q.," http://www.icao.int/safety/CMAForum/Pages/FAQ.aspx.

    c. Conflict avoidance and resolution—When to speak up and when to shut up. When arguments and confrontation with airline staff and crew or other passengers can take an ugly turn for the worse.

Car rentals

1. Clearly specify all preferred suppliers and corresponding corporate discount numbers (mandatory use), including:
   a. Where each supplier should be used.
   b. Which rates include insurance, and where coverage applies (including exclusions).
   c. Recommend reconfirmation of insurance in rate before leaving rental lot.
      i. Ensure that corporate discount number is on rental agreement if it is required for insurance coverage.
   d. Specify approved car sizes and types allowed, which are covered by supplier-provided insurance.
2. Restricted use of car rentals after transcontinental or international redeye flights.
3. Clear instructions for car rentals in locations where preferred supplier or corporate discount and insurance is not available (i.e. purchase of full coverage, use of corporate credit cards, etc.)
4. Use or reliance of personal auto coverage is prohibited for business car rentals
5. Secondary or additional authorized drivers on business travel car rentals is prohibited unless the additional driver is also on business travel paid for by the company.

Rail travel

Worldwide

Many millions of passengers use suburban and commuter railways around the world each day. Some of the daily passenger numbers are staggering, such as Sao Paolo, Brazil's Companhia Paulista de Trens Metropolitanos (CPTM), carrying more than 3 million passengers daily, or Mumbai, India's Mumbai Suburban Railway, which carries approximately 6.9 million passengers per day.[4] Most Westerners think of Europe or the United States when it comes to rail travel as a common mode of transportation, but rail travel is one of the most cost-effective, widespread, low-cost transport systems on the planet. Unfortunately, there are no global standards for railway safety standards. Some countries and regions have standards, but we are yet to see the kinds of security that we now expect at airports around the world, evenly facilitated at railway stations globally. In the United States, Amtrak's website's safety and security page says that Amtrak only conducts random screening and inspection of passengers and their personal items.[5] This clearly shows rail travel's vulnerability in terms of terrorism-related violence. The statistics provided below for rail-related accidents and fatalities do not take into account terrorism, as to date rail travel has seemingly not been a primary focus for terrorist groups. However, the 2004 Madrid train bombings that killed 191 people, and the

---

[4] Wikipedia, "List of Suburban and Commuter Rail Systems," https://en.wikipedia.org/wiki/List_of_suburban_and_commuter_rail_systems.

[5] Amtrak, "Safety & Security," http://www.amtrak.com/safety-security.

attempted attack in France in 2015that was thwarted by three Americans[6] are reminders that all passengers should be aware and vigilant in practicing personal awareness and safety best practices when traveling via rail, and that we need increased passenger and baggage screening, similar to those used at airports. Regarding the August 2015 incident in France, it is shocking that a passenger could board a train with an AK-47. If that's the case, how much easier would it be for anyone to bring smaller guns on board? Accident statistics aside, the vulnerability for acts of terrorism on passenger trains worldwide, should give both employers and travelers extreme concern over personal safety.

For a fascinating glimpse of some of the worlds suburban and commuter railway passenger data, visit https://en.wikipedia.org/wiki/List_of_suburban_and_commuter_rail_systems.

United States

According to the Federal Railroad Administration and the Bureau of Transportation Statistics in the United States, during the 25-year period from 1990 until just after the 2015 Philadelphia Amtrak derailment, there have been a total of 213 passenger deaths.

According to a 2013 study[7] in the journal *Research in Transportation Economics,* commercial air transportation was the safest mode of transport with 0.07 fatalities per billion passenger miles. The study period covered the years 2000 to 2009 and excluded acts of suicide or terrorism. Next after air travel, was bus transportation with 0.11 fatalities per billion passenger miles, followed by long-haul passenger trains at 0.43 fatalities per billion passenger miles. Automobile accidents by far are still the most dangerous mode of transport with a 7.3-per-billion-passenger-miles rating. So, statistically in the United States, although not as safe as commercial air travel, rail travel is still far safer than traveling by car.

Europe

Each year the European Railway Agency (ERA) publishes a document called "Railway Safety Performance in the European Union,"[8] which contains details and statistics on rail transportation within the EU member countries. The ERA is the foundation for the EU's strategy for railway safety, supporting national investigation bodies (NIBs) and national safety authorities (NSAs) with their tasks, and providing evidence at the EU level for policy actions. The ERA provides and promotes a common safety framework across the EU to achieve an open railway market, and supports the European Commission on developing EU legislation.

---

[6] CNN, "Spain Train Bombings Fast Facts," March 11, 2015, http://www.cnn.com/2013/11/04/world/europe/spain-train-bombings-fast-facts/.
[7] Ian Savage, "Comparing the Fatality Risks in United States Transportation Across Modes and Over Time," *Research in Transportation Economics*, 2013;43(1):9–22, available at http://www.sciencedirect.com/science/article/pii/S0739885912002156.
[8] http://www.era.europa.eu/Document-Register/Documents/SPR2014.pdf

Important European Commission Railway Accident Definitions[9]

Significant accident–directive 2004/49/EC, commission directive 2009/149/EC and regulation (EC) no. 91/2003

"Significant Accident" means any accident involving at least one rail vehicle in motion, resulting in at least one killed or seriously injured person, or in significant damage to stock, track, other installations or environment, or extensive disruptions to traffic. Accidents in workshops, warehouses and depots are excluded. Significant damage is damage that is equivalent to €150,000 or more.

Serious accident–directive 2004/49/EC

"Serious Accident" means any train collision or derailment of trains, resulting in the death of at least one person or serious injuries to five or more persons or extensive damage to rolling stock, the infrastructure or the environment, and any other similar accident with an obvious impact on railway safety regulation or the management of safety; "extensive damage" means damage the can immediately be assessed by the investigating body to cost at least €2 million in total.

Every year more than 2000 significant accidents occur involving EU Member States' railways, accounting for an estimated cost of up to €1.7 billion . According to the "Railway Safety Performance in the European Union" report for 2014, their latest safety indicators data states that railway safety continued a progression of improvement across the EU in 2012 with 2068 significant accidents, which included 1133 fatalities and 1016 people seriously injured (representing a 7 percent decrease in the number of significant accidents and a 5 percent decrease in casualties as compared to 2011). In 2013, there were five qualifying serious accidents in the EU involving train collisions, derailments, and other accidents.

Considering that rail travel is much more common in the EU than in the United States, it stands to reason that EU statistics are different and that rate rail travel is safer than bus travel, falling just behind commercial air transportation. Using the ERA's statistics over a different period of time than was previously quoted for the United States, between 2007 and 2012 and using a measurement of fatality risk per billion passenger kilometers (not miles), the EU has a lower risk rating of 0.13 fatality risk per billion passenger kilometers versus the United States at 0.26 and Canada at 0.14. The number and types of incidents included in the ERA's report vary by country, but the report provides detailed numbers of incidents per country that can help companies make better judgments about the level of risk involved with rail transportation throughout the EU.

Hotels

Educated travel managers are moving beyond preferred hotel property selection based solely on price and perks. Savvy buyers are aggressively incorporating increasing

[9] http://www.era.europa.eu/Document-Register/Documents/SPR2014.pdf

safety standards into their annual hotel request-for-proposal (RFP) programs. Because it can be a challenge to obtain validated answers to a long list of safety questions from individual hotel properties, which is why buyers should consider the following levels of assessment and review for their corporate hotel programs:

1. Adopt developing industry standards, such as Sweden-based Safe Hotels Alliance, which uses a standardized process for certifying and maintaining hotel safety criteria for participating properties.
2. Adopt a minimum standard of questions to ask relative to safety and security that may not already be covered by your hotel RFP template. (See examples from Rezidor Hotel's "Always Care" program details in Chapter 7.)
3. Any extensive hotel safety and security requirements may be more easily certified in partnership with global hotel chains or networks that can spend the time and money to ensure that buyer standards are adhered to, and standards may be communicated via a designated point of contact for all participating properties within the group in question.
4. On an ad hoc basis, pay a third party to conduct hotel risk assessments (either annually, or as needed for long-term projects, etc.).

Buyers should pay close attention to those suppliers that are making serious investments into developing safety and security standards. These are investments that are more than simply checking a box on a form; instead, the suppliers are adopting real programs that are implemented and made practical use of, along with continuous process improvement.

Ground transportation

With mobile technology making its mark on the business travel market, the ground transportation and black car services industries have not been spared its influence on changing their business models. Mobile technologies have enabled new service models for car sharing services, including use of private assets by contractor-suppliers for transportation services that aren't necessarily black car services, but are not quite taxis either. Black car services and taxis still exist, but are being forced to introduce technology similar to the new "sharing economy," or peer-to-peer, like suppliers such as Uber and Lyft. These new services have introduced a litany of security concerns to business travel buyers and travelers, and suppliers are working hard to provide assurances that safety is being addressed and that buyer concerns are being heard.

With the introduction of so many new suppliers in recent years to the ground transportation market, each trying to differentiate its approach from that if its competitors, understanding the differences can become confusing. A high-level breakdown of the differences in services are as follows:

Taxicabs

Using New York City as an example, taxicabs, as defined by city law, are the only car services allowed to pick up passengers (fewer than nine, excluding the driver) that hail them from the street or at designated taxi stands. In New York City, taxicabs are a metered fare, clearly colored yellow, and marked with roof lights indicating if they are for hire, currently hired, or off duty.

For-hire vehicles

Using New York City as an example, for hire vehicles (FHVs) are prearranged services using vehicles that accommodate fewer than nine passengers (excluding the driver), which are regulated by local FHV organizations that handle complaints or customer service issues. FHV companies can include car services, black cars, and limousines. Car services are a type of FHV that typically serves local or neighborhood markets. Black car services are FHV services that primarily serve business travel clients using luxury vehicles. Limousines are also a type of FHV services, typically charging based upon "garage to garage" drive times in luxury vehicles, including, but not limited to, stretch limousines (Lincoln Towncars, Cadillacs, etc.)

"Sharing economy" ground transportation suppliers

The "sharing economy" or "shared economy" is, in essence, collaborative consumption of excess capacity in goods and/or services (mostly privately owned by drivers for ground transportation), enabled via an online marketplace or platform that facilitates transactions between customers and drivers. Some of the most common examples discussed by business travel buyers today are Uber and Lyft, which are listed here for discussion purposes specifically as sharing economy for ground transportation.

IMPORTANT: These companies are not transportation service providers. They can be best described as technology application service providers, and their terms and conditions convey this, and also very specifically what they are not liable for with the acceptance of their terms and conditions, which users automatically agree to if using their applications or platforms.

Questions to ask when considering shared economy ground transportation suppliers for business travel include:

1. Are companies like sharing-economy ground transportation providers allowed per policy?
2. Does policy designate which of these services are allowed for business travel use, and which (if any) are prohibited?
3. If use of such suppliers is allowed, has your company verified that the providers in question meet safety, security, and insurance requirements?
4. What is the company's policy for booking and reimbursement of such services?
   a. Is it restricted to specific locations, or wherever available?
   b. Are these services being expensed as "taxis" or another expense category?
5. Does the company provide any special training for use of shared economy suppliers?
6. What kind of safety training do these suppliers require of their drivers, and how often do the drivers need to take it?
7. Do these suppliers require regular drug testing?
8. How often are maintenance checks done of vehicles, and who is conducting these checks?
   a. What are the maintenance standards (tune-ups, brake replacement, oil changes, etc.)?
   b. Details on Lyft's position on maintenance inspections can be found at https://help.lyft.com/hc/en-us/articles/214219517-The-Car-Inspection. However, having a "mentor" driver in some cirumstances conducting the inspections may make some employers nervous in consideration of whether or not the mentor is qualified to evaluate the mechanical stability and safety conditions of the vehicle adequately.
   c. Lyft's vehicle safety requirements can be found here: https://help.lyft.com/hc/en-us/articles/214219557.

9. Are the drivers are considered contractors or employees? What are the legal ramifications of that designation?

10. Some sharing economy ground transportation providers state that their US$1 million third-party liability coverage is primary coverage, which kicks in from the moment the driver accepts the ride until the ride is concluded. However, if an accident is catastrophic with life-changing disabilities requiring long-term care, especially if the accident involves more than one passenger, is this enough coverage?

11. How can the platform service provider's gap or umbrella insurance coverage apply when agreement to their terms of use for using their platform appears to infer that the user indemnifies them from liability?

12. What are the different insurance coverage or exceptions to coverage, based upon the type of product being used (some suppliers offer multiple services)? Uber's website states that "As always, all UberBLACK, UberSUV, or uberTAXI rides are provided by commercially licensed and insured partners and drivers. Those transportation providers are covered by commercial insurance policies, in accordance with local and state requirements. We are proud of these policies."

   However, with so many variables by jurisdiction and product type, understanding what levels of coverage a business traveler has (at the employer and the traveler level), can be confusing. A network standard would be ideal, even if it provided more coverage than required in a particular jurisdiction.

13. These same questions apply to Lyft and its positions, which may be interpreted as similar to those of Uber.

14. What effect do the laws of the jurisdiction where the service is to be provided have on insurance coverage or exceptions to coverage and on waiver of liability clauses?

Details regarding Uber's insurance coverage for UberX services can be found at http://newsroom.uber.com/2014/02/insurance-for-uberx-with-ridesharing/. Lyft's insurance coverage can be found at https://help.lyft.com/hc/en-us/articles/213584308-Insurance-Policy.

Training - As of mid-2015, Uber's YouTube driver training video was approximately 13 minutes and 36 seconds long, and vehicle maintenance processes and standards are difficult to identify on a broad scale, except for some details about squeaky brakes, body vehicle body damage, broken air conditioning, and missing hubcaps.

Uber also has a public safety video on YouTube called "Safety By Design," which is 1 minute and 31 seconds in length. The video includes rider testimonials about the driver and rider-rating process, and riders having digital records of their experience with Uber, and details on the Uber background check process.[10] However, at the time of writing this book, Uber recently continued to struggle with the accuracy and effectiveness of its background checks.[11] Uber continues to dispute the value of using fingerprints in their background check process (see: http://www.boston.com/business/2016/02/23/uber-advisor-doubles-down-fingerprint-opposition/IsN6awAopOho1WiaQ9wR5K/story.html). However, their current background check process only goes back 7 years, and while fingerprint usage and technology aren't always 100% accurate, they could be used to go back further than 7 years and if they

---

[10] Uber, "Safety By Design" (video), March 25, 2015, https://www.youtube.com/watch?v=lcYpoXn3L3k.

[11] http://www.latimes.com/local/lanow/la-me-ln-uber-background-checks-20150819-story.html.

increase the number of convicted felons rejected from their program, even a small increase would be an improvement.

Some details regarding Uber's driver requirements and screening process, can be found at http://newsroom.uber.com/2015/07/details-on-safety/. At this link, you can find Uber's approach to screening standards for different Uber services. Details on Lyft's driver requirements can be found at https://help.lyft.com/hc/en-us/articles/213585758.

Uber does not allow hailing of drivers, provides for anonymous feedback, and for each confirmed ride, its application users can see a driver profile that includes the driver's name, license plate number, photo, and rating. However, there have allegedly been reports of drivers using different cars than what is identified in the order, but allegedly it is often because the driver didn't adjust their profile, when in fact the driver may use two cars and have both registered. It's important that riders pay attention to the car type, license plate, and driver photo, to make sure that it all matches what is in the order and to refuse a ride if one or more of those things don't match.

An excerpt from Uber's Terms and Conditions includes the following:

YOU ACKNOWLEDGE THAT UBER DOES NOT PROVIDE TRANSPORTATION OR LOGISTICS SERVICES OR FUNCTION AS A TRANSPORTATION CARRIER.
UBER DOES NOT GUARANTEE THE QUALITY, SUITABILITY, SAFETY OR ABILITY OF THIRD PARTY PROVIDERS.[12]

An excerpt from Lyft's terms and conditions includes the following:

LYFT DOES NOT PROVIDE TRANSPORTATION SERVICES, AND LYFT IS NOT A TRANSPORTATION CARRIER.
EACH TRANSPORTATION SERVICE PROVIDED BY A DRIVER TO A RIDER SHALL CONSTITUTE A SEPARATE AGREEMENT BETWEEN SUCH PERSONS.[13]

Both Uber and Lyft's terms and conditions have language that requires an indemnification of the company and that deems the terms and conditions as accepted by using their technology platforms.

Details on Lyft's approach to safety, including their US$1 million insurance coverage (also listed as primary coverage to the driver's insurance), can be found at https://www.lyft.com/safety.

Safety concerns over shared economy ground transportation

In addition to concerns over insurance coverage, the following items should be considered with regard to safety when using sharing economy ground transportation:

1. Aside from the amount of insurance provided by the platform provider, understanding when a driver's personal or commercial insurance or the technology platform provider's coverage is applicable can be confusing. In the event of an incident where more than one party's

---

[12] Uber, Legal, "Terms and Conditions" for the United States, April 8, 2015, https://www.uber.com/legal/usa/terms.
[13] Lyft, "Terms of Service," November 3, 2015, https://www.lyft.com/terms.

insurance involved (e.g., the driver and the platform service provider), how seamless is the claims process, and is it handled by one party or many?

2. Sharing economy drivers aren't always subject to the same transportation authority regulations as taxis, and often are not required to have comprehensive safety training courses (excluding unsupervised online training videos as used by some sharing economy suppliers).

3. Uber's partner safety video (as of May 2015) lists requirements for operating the partner's vehicle, such as working seat belts, and the cars must be in "great working condition." According to Uber, if anything is faulty with a vehicle, such as the "lights, brakes, windshield wipers, or tires," the drivers shouldn't operate the vehicle. (Questions: What are the definitions or standards for "great working condition"? Who inspects vehicles to ensure the vehicles meet minimum safety standards? Why can't these inspections and maintenance inspections and standards be better defined and standardized across multiple states and countries? Even if that would require setting the standard higher than required for some locations, the customer base would at least understand the standard from one country or state to the next for the service purchased, even though the standards may still be different for other products because of commercial insurance requirements for more premium services like UberBlack.) Using terms like "functional' or "good working condition" are subjective, especially if being inspected by someone who may not be a qualified mechanic.

4. Beyond background checks, are there any advance, validated (via inspection) safety requirements for drivers prior to their being able to provide services?

    a. In response to the California Public Utilities Commission's mandate to address public safety via required driver training programs, a ride-service subsidiary of Uber (Rasier) responded as follows: "Rasier will also recommend that partners with less than five years of driving history take a driver training course at a school listed in the California DMV's Occupational License Status Information System database. For each California city where Rasier partners operate, Rasier will provide, during the application or onboarding process, a list of three such local driving schools."[14] (Note the use of the word "recommend" versus "require" or "mandate." Also, what about drivers with more than five years of experience prior to service and then on an ongoing basis?)

Some example policy considerations for sharing economy or traditional ground transportation services include:

1. Use of unmarked, unlicensed, nonregulated car services offered to passengers hired upon arrival at destinations is strictly prohibited ("gypsy cabs").

2. When using a vetted, preferred supplier car service, arrange for the name and photo of the driver in advance (even if via e-mail or text prior to arrival from dispatch only). Many services offer vehicle and driver identification via mobile applications, and some even show the location of the car in proximity to the traveler while waiting on the intended ride.

3. Require all car services that are providing any "meet-and-greet" services to only list security program code names, on arrival placards. Never list the individual traveler's name or company name.

In summary, in order for sharing economy ground transportation providers to improve their standards for insurance, driver background checks, driver training and vehicle maintenance standards (with qualified mechanics), the market is going to have to demand it across these supplier networks. This could conversely drive better

---

[14] Jon Brooks, "What UberX Drivers Are Saying About Their Training and Safety Issues," KQED News, February 20, 2014, http://ww2.kqed.org/news/2014/01/30/lyft-uberx-driver-safety.

standards for the taxi industry as well. In consideration of the Feb. 2016 shootings in Kalamazoo, Michigan by an Uber driver with no criminal record, employers and travelers are going to have to ask themselves "What is it going to take in terms of some new market standards to feel safer and for my company to embrace these suppliers via policy?" Should the Kalamazoo driver even been allowed to carry a firearm in the car, even if he had a permit to conceal and carry? If he would theoretically be allowed to carry a firearm in the car, how comfortable are riders getting into a car with someone carrying a gun? Would they want to know that?

"Sharing economy" accommodations providers

As mentioned earlier in this chapter under "Ground Transportation," "sharing economy" suppliers include ground transportation, but this emerging market also includes accommodations. This section discusses the pros and cons of the sharing economy accommodation providers.

Like ground transportation, these providers, such as Airbnb and HomeAway, are not hotel or hospitality companies, but are technology platforms that enable users to connect with host owners of private accommodations for personal or business travel usage.

Many of the providers offer the entire condo, house, or apartment for rent, but some "hosts" also rent a room in their homes (while still occupied by the hosts or other guests), as well as tents or camping spots on private land.

Many travel buyers refer to these kinds of suppliers as "disruptive" to the market, as they are "shaking things up a bit," providing more inventory around the world and bringing prices typically down for accommodations. While these are valid points, travel buyers are taking a closer look at safety related concerns with sharing economy accommodations providers, which also prompts them to change the way that they look at their traditional hotel suppliers in terms of risk and safety standards (see "Rezidor Hotel Group Safety and Security 'Always Care' Program Case Study" in Chapter 7).

Companies like Airbnb and VRBO are similar to Uber and Lyft in the sense that they act as a facilitator or technology platform that connects users selling time at or usage of their properties with users who want to purchase said time or usage. These venues or platforms claim to not be a party to the agreements between users, so bear that in mind, along with the previous statement that these providers are not hoteliers or hospitality companies.

Key considerations for use of sharing economy accommodations suppliers include, but are not limited to, the following:

1. Mandatory safety standards for all properties
   a. Fire
      i. Are hosts required to provide proper fire escape routes to all guests (not just recommended)?
      ii. Are there usable, unexpired fire extinguishers in the unit, with disclosures on where they are stored and how to use them?
      iii. Are there ample fire and carbon monoxide alarms installed?
      iv. Are hosts required to provide local fire department numbers to guests upon arrival?

  b. Sanitation

    i. Are all soaps, toiletries, and linens changed between guests?

    ii. Are all bathrooms and kitchens sanitized between guests?

    iii. How often, if at all, are the beds and furniture checked for bed bugs?

    iv. If only renting a room in a host-occupied facility, what are the sanitation risks involved based upon host hygiene and personal health issues?

  c. Security

    i. Is it possible that multiple people or previous guests have keys to the unit in question during a traveler's stay?

    ii. Are hosts required to provide local police department numbers to guests upon arrival?

    iii. If only renting a room in a host-occupied facility, what are the security risks involved with sharing the space with someone based upon their mental health and criminal history?

2. Conflict resolution with provider or host

3. Insurance (theft, bodily injury, intellectual property)

4. Access to emergency services

5. Host or provider background checks or screening, and for what types of crimes? (e.g., sex offenders, violent crime?)

An excerpt from Airbnb's Terms and Conditions includes the following[15]:

PLEASE NOTE THAT, AS STATED ABOVE, THE SITE, APPLICATION AND SERVICES ARE INTENDED TO BE USED TO FACILITATE HOSTS AND GUESTS CONNECTING AND BOOKING ACCOMMODATIONS DIRECTLY WITH EACH OTHER. AIRBNB CANNOT AND DOES NOT CONTROL THE CONTENT CONTAINED IN ANY LISTINGS AND THE CONDITION, LEGALITY OR SUITABILITY OF ANY ACCOMMODATIONS. AIRBNB IS NOT RESPONSIBLE FOR AND DISCLAIMS ANY AND ALL LIABILITY RELATED TO ANY AND ALL LISTINGS AND ACCOMMODATIONS. ACCORDINGLY, ANY BOOKINGS WILL BE MADE OR ACCEPTED AT THE MEMBER'S OWN RISK.

Regarding background checks for hosts or guests, as of August 22, 2015, under its website's "Updated Terms of Service," "Terms of Service," "Disclaimers,"[16] Airbnb states:

YOU ACKNOWLEDGE AND AGREE THAT AIRBNB DOES NOT HAVE AN OBLIGATION TO CONDUCT BACKGROUND OR REGISTERED SEX OFFENDER CHECKS ON ANY MEMBER, INCLUDING, BUT NOT LIMITED TO, GUESTS AND HOSTS, BUT MAY CONDUCT SUCH BACKGROUND OR REGISTERED SEX OFFENDER CHECKS IN ITS SOLE DISCRETION.

[15] https://www.airbnb.com/terms
[16] https://www.airbnb.com/terms

Airbnb Case Study

A 19-year-old Boston man travels to Madrid, Spain, and books accommodation via Airbnb. Upon arrival, the host locked the man in the apartment and removed the key. With the host still present and rattling knives around in the kitchen, they pressured him for sex. The young man began messaging his mother for help. A call by his mother to Airbnb resulted in their refusing to provide the address or to call the police, but instead advising her to call the Madrid police and ask them to call them for the information. However, the number that she had for the Madrid police answered with a recording in Spanish that would then disconnect her, and then her additional calls back to her Airbnb contact would go straight into voicemail.

Now mind you that providing such an address over the phone to a third party is a security risk, and regular hotels will neither confirm nor deny if a specific person is staying at their hotel upon an inquiry from someone via phone, but hotels will often send security to investigate if there is a potential or suspected problem. With sharing economy accommodation suppliers, employers and travelers need to consider what kind of safety and security support is available to them. Even in the absence of formal programs provided by a supplier, this is a very good example of why all travelers need access to a global crisis response hotline via a professional TRM services firm and that any business travel data (such as a sharing economy accommodation address) should absolutely be captured and recorded properly in the employer's TRM traveler tracking platform.

Eventually, the young man convinced his captor that he was meeting friends who knew where he was staying and was allowed to leave with his belongings, thereafter returning home to undertake trauma therapy.

In a *New York Times* article covering this story,[17] an Airbnb advisory board member appallingly stated that the traveler or travelers should ask, "Is there a deadbolt that only I can turn," or at least look to see if someone could lock you in. While Airbnb's terms explicitly indemnifies it from the actions of its platform users, there are still many security issues that Airbnb need to address in order to gain the business travel buyer community's confidence.

---

[17] Ron Lieber, "Airbnb Horror Story Points to Need for Precautions," *The New York Times*, August 14, 2015, http://mobile.nytimes.com/2015/08/15/your-money/airbnb-horror-story-points-to-need-for-precautions.html?_r=0&referrer=.

Crisis response

5

This chapter discusses what it means to have your own internal crisis response plan, and the key components of one, as well as how it works in concert with third-party resources to manage incidents as they arise.

What constitutes a crisis within the framework of travel risk management?

Although the answer may vary from company to company, the general definition is that a crisis is "an incident with the potential for severe bodily injury, loss of life, or mass travel delay." This definition is very subjective and is broadly used on an industry level. On an organizational level, depending upon a company's culture and risk threshold, a crisis can be defined as an incident that impacts a distressed individual traveler, a loss of company property with the potential for intellectual property loss, gross misconduct of an employee or traveler that could impact the company's reputation, etc. One might consider natural disasters or terrorism in the context of travel-related crises. However, depending upon how one views a situation, a company could easily consider a senior-level executive being arrested for solicitation of prostitution in another country while on business a crisis. The potential damage that such arrest could do to the company's image should the incident be leaked to the press could be difficult to manage. Another example might include the inability to get key personnel into facilities or critical business meetings because of third-party or environmental interference, such as border closures as a result of medical quarantines or civil unrest.

Again, each company's approach to risk is different, so any sample program outlines may vary from one organization to the next, but consider the following components when creating a crisis response plan for your company:

- Create crisis team(s)
- Designate team leadership
- Business continuity plans
  - Business impact analysis by incident
  - Eminent disruption/contingency plans
  - Pandemic plans
- Contract third-party crisis response support
- Define crisis response protocols, thresholds, and escalations
- Designate organizational command center
- Define organizational crisis communications process
- Define reporting and status process
- Conduct crisis "readiness' exercises
- Continuous process improvement
- Insurance and payment options for support services and case management

Crisis management team

First and foremost, companies need to designate an internal crisis management team that includes senior-level executives who are authorized to make final decisions as they are related to risk.

- Designate team members capable of making strategic decisions as needed when called upon, including hierarchy of decision making authority.
- Document each team member's complete contact information, including afterhours information (always include mobile phone numbers). Make this available to contracted crisis response supplier.
- Ensure that alternates are in place for each team member in case of a planned or unplanned absence of a primary team member.
- Designate team leadership by protocol, region, or task.
- Designate information technology (IT) team leader and support staff contacts in case of intellectual property theft or loss, or hacking of company computer or mobile phone.
- Human resources must be represented on the team for employer and employee labor law and regulatory compliance and human capital support.
- Legal resources must be represented on the team for incidents involving violations of law between traveler and company or traveler and third parties while representing the company.
- Facilities resources must be represented on the team for utilization of facility resources and supplies, closest resources to facilities, and security effectiveness of facilities to shelter in place when necessary.
- Marketing and communications resources must be represented on the team to provide information and announcements, both internally and externally (including the media, law enforcement or regulators when necessary).

Travel risk management and crisis management should not exclusively be motivated by money or the bottom line, but because it is the "right thing to do." However, when facing a committee or board that isn't swayed by moral obligations or legal precedent based upon duty of care, showing the committee or board the Accenture research document "Markets Don't Lie. Tracking a Crisis Through Share Price,"[1] which shows a correlation between a company's share price following a crisis and should hopefully, motivate stakeholders to be more proactive.

Risk disclosures and crisis response

1. Third-party, nongovernmentally sponsored safety and risk intelligence must be made available to all travelers and explained within the policy as to how and when this information is to be made available, such as:
   a. Link to third-party database via company intranet
   b. Links to third-party database via traveler itineraries
   c. Push of alerts and risk intelligence via e-mail based upon itinerary destinations (*always* provide risk disclosures to travelers via *both* push and pull communications, documenting provision of and access to information within policy)

[1] http://spotidoc.com/doc/169011/corporate-crisis-management

2. Document emergency medical services and insurance provider details within policy, outlining coverage and exclusions where possible (include in hotline protocols)
3. Provide a single phone number for all crisis response support to travelers for all medical and nonmedical emergencies, including, but not limited to:
   a. Medical emergencies and evacuations
   b. Personal safety emergencies and evacuations
   c. Loss of property or intellectual property

Business continuity plans

Even though companies spend considerable time and effort on their business continuity plans, there are several components of a business continuity plan related to risk that should always be added and adopted as industry standards, addressing many "what if" scenarios that can impact your business. Additions to be considered include:

1. Business impact analysis—What if the only international airport near your offices closes down or is shut down for some reason? What if your employees can't leave the country because the area between your facility and the airport is under quarantine or the roads have been closed? What if there is civil unrest near or around your offices or wherever your key stakeholders are traveling to, and your people can't leave the office or can't get to the office safely for several days? What is the impact of these various scenarios on your business? Companies should use a business impact analysis to help you determine what the potential losses or costs to the business could be in the event of general or sometimes known specific critical incidents. Examples to use for business impact analysis might be:
   a. Natural disasters
   b. Transportation disruptions
   c. Civil unrest
   d. Biohazards and pandemics
      Establishing some metrics for potential costs to the company because of the loss of productivity per day for individuals as well as corporate facilities will help with the values associated with significant interruptions.
2. Eminent disruption plans—What if access to your facilities was interrupted for a long period of time? What if all standard communications were down for an extended period of time at and near our facilities? What if access to water and electricity were unavailable? Similar to a business impact analysis, eminent disruption plans are what appropriations and/or arrangements have been planned for under the circumstances where travel or operations have been disrupted without much notice. These plans should always be addressed in layers or levels, with specific goals in mind at each level, starting with:
   a. Risk assessment process before and after disruptions
   b. Identification of regulations that establish minimum requirements for your plans
   c. Security and safety of all stakeholders (personnel, contractors, visitors, locally impacted environment at large)
      i. Transportation
      ii. Emergency response (fire, ambulance, etc.)
      iii. Shelter in place
      iv. Employee assistance

    d. Increasing supply levels of things like emergency food, water and medical supplies

    e. Alternative means of communications (hard line connections, satellite phones, Internet connectivity, etc.)

    f. Minimal operational functionality—define what is minimal functionality, from the individual to the facility

    Additional areas to address in eminent disruption plans are potential violations of the FCPA (U. S. Foreign Corrupt Practices Act), particularly for publicly traded companies and their officers, directors, employees, stockholders, and agents, and other legal and ethical issues as a result of your plans and decisions.

3. Biohazard/Pandemic Plans—Whether from internal or external exposures to chemicals, viruses, bacteria, or other harmful substances, every business continuity plan must have a supplemental biohazard/pandemic plan. Using the World Health Organization's pandemic plan checklist for influenza as an example, some high-level points to consider in your biohazard/pandemic plans are:

    a. Preparing for an emergency

        i. Command and control

        ii. Risk assessments

        iii. Communication (internal and external)

        iv. Legal and ethical issues

        v. Response plan by pandemic phase

    b. Surveillance

    c. Case investigation and treatment

    d. Preventing spread of hazard or disease

Defining contracted third-party crisis response support

1. First line of support for travelers—One hotline number to call globally for any kind of perceived emergency (excludes general travel reservations and changes, unless commercial travel is unavailable or traveler needs additional assistance in facilitating safe travel).

2. Case management and documentation of all incidents—Having an intermediary for collective recording and documentation of all interaction between impacted parties, company stakeholders, suppliers, insurance providers, etc., via the designated hotline, to ensure that company policies and protocols are being adhered to, from a cost, legal, ethical, moral and regulatory compliance perspective.

3. Recording of all calls.

4. Facilitation of company's predefined crisis protocols (whom to call/when). When to handle situations that are preauthorized, for example, medical services under specified spending limits (e.g., based upon established protocols, providing automatic approval for payment of non-life threatening treatment under $50,000 USD (amount set by company protocol and/or insurance policies). Approval may be in conjunction with hotline, case management representative coordinating with the company's designated insurer and medical service providers for the region of the world where the traveler needs treatment.)

5. Use of third-party resources (extraction, secure ground transport, executive protection, medical support and/or evacuation, etc.).

6. Coordination with client insurance providers or suppliers for service orders or payment of services.

7. Active traveler monitoring via tracking and outreach as directed by company policies and protocols.
8. Provision of alert and various other forms of risk intelligence, which is useful in mitigating and managing crisis response (both proactive and reactive disclosures).

Define crisis response protocols, thresholds, and escalations

Before you can fully define your crisis response protocols, organizations must clearly define who is in charge in a crisis situation. Does the decision authority change by location or type of incident? As with every aspect of risk management planning, it begins with assessments.

Essential assessments in crisis response protocol planning include:

1. Does your insurance or contracted emergency medical support and evacuations services provider cover your travelers globally?
   a. Understand the coverage and its limitations
      i. Will it guarantee third-party payment for services, or must the individual or company provide payment and file for reimbursement later?
      ii. What are the processes required by the insurers or service providers to qualify for coverage and payment?
      iii. What exclusions or gaps apply to your coverage, if any?
2. Does your crisis response support, provide nonmedical security resources globally?
   a. Are these services (e.g., nonmedical evacuations, secure transport, executive protection) covered by insurance?
      i. Who provides the coverage?
      ii. What is covered and where?
      iii. What are the processes required by the insurers or service providers to qualify for coverage and payment?
      iv. What exclusions or gaps apply to coverage, if any?
3. Are predefined authorizations for services set up with your crisis response support for costs up to a specific amount?
   a. Who is designated to authorize payment for costs over these amounts (by incident type, region, or amount)?
4. How do policies impact crisis response protocols?
   a. Maximum employees per flight
   b. Prohibited use of certain suppliers
5. What processes or types of incidents trigger activation of corporate command center or crisis task forces (e.g., anything that threatens business continuity)?

Bear in mind that crisis response protocols are triggered by an incident already having taken place. Unlike pretrip or avoidance processes or protocols, crisis response and mitigation protocols are "post incident." Some general protocol categories include:

1. Medical-related support calls (nonevacuation)—Transfer and support of calls from travelers needing medical support, but not requiring evacuation, to client-contracted medical support providers/insurers. This usually does not require crisis response team involvement, unless

support exceeds an expense threshold above what insurance covers and/or policy sets. These incidents are often completely handled by third-party crisis support, with reports being sent to company stakeholders at preset intervals.

2. Medical-related support calls (with evacuation)—Transfer and support of calls from travelers needing medical support that leads to a necessary medical evacuation to either the traveler's home country or the nearest equipped medical facility where the most appropriate support for the patient's needs can be given in a timely fashion. This usually involves the notification of and communication with a crisis response team member, according to policy, especially if any additional guidance or approvals are needed.

3. Security-related support calls (nonevacuation)—These calls can range from fear regarding personal safety because of a stalker or civil unrest, or problems with the local authorities, and often are resolved by the third-party crisis response centers, by counseling via phone and/or engaging their local contacts near the traveler to provide support on an as-needed basis. Unless the incident involves the potential for excess liability or expenses exceeding amounts preauthorized for the third party to manage, a crisis response team would not likely be notified until receiving a copy of the reports documenting the incident. Within this category, companies would likely document with third-party crisis response support providers, things like who their insurance providers are for different types of security issues in different regions of the world, such as kidnap and ransom insurers and support, or if no insurers are on file for certain incidents, such as nonmedical evacuations, whether the company will pay directly for the cost of services rendered. What amounts are preauthorized, before requiring the involvement of a crisis response team member and from what region, are also documented.

Risk protocol thresholds can be related to pre-authorized expense amounts based upon things such as risk ratings for a location, which is a good example of how protocols for different incidents and locations may vary (e.g., pre-authorization of armed and guarded ground transport may be part of a standard protocol for travelers in Egypt, but not in the UK, based upon different location risk ratings and other factors).

Examples of some risk thresholds incorporated into crisis response plans may include the exclusion of travel to or via countries with a level 3 or above risk rating on a scale of 1 to 5. Another example could be the qualification that payment for extraction or nonmedical evacuation not covered by insurance must meet predefined criteria. The difficult truth about crisis response is that sometimes employers often set monetary limits for certain types of actions. It's just the reality of things, but at least when these limits are set, the employer has thought through these different use cases and is more prepared than those employers who have no policies or protocols in place in the event that internal or third-party crisis response and support is required.

Designate organizational command center

Generally, senior business continuity stakeholders are a part of the organizational command center, responsible for brainstorming solutions and taking part in crisis response exercises based upon policies and procedures that they create. Even so, you still need a chain of command in which someone is in charge and can make costly and/or critical decisions on behalf of the company. Regardless of the decisions made during a crisis by delegates, the person in charge should always be informed, as delegates may have been acting in her/his absence, or were authorized to act by the primary stakeholder.

There should be a primary meeting place designated for the group, documented with address and phone information, along with an established conference line and/or web conference capabilities, as well as a central command center e-mail address as an alternative form of communication from the field for status updates, if other methods of communication are down. Separate from a global command center, regional command centers should be considered as well, based upon the concentration of business facilities and assets in the respective regions. The use of a command center, whether physical or virtual, is typically reserved for incidents of extreme circumstances that your third-party crisis support cannot handle alone, based upon your predefined policies, insurance, and established risk-and-spend thresholds.

Included in company policies and procedures should be a process for what triggers the activation of one or more command centers (based upon type or severity of the incident in question), and a method for notifying command center stakeholders, requesting when and where they should assemble. Once assembled, in conjunction with your third-party crisis response provider's support, stakeholders should:

- Assess the situation based upon intelligence provided by their third-party support and any other appropriate resources
- Log/document the situation as much as possible
- Assign tasks to command center stakeholders
- Deploy resources
- Monitor status with continued documentation
- Provide executive briefings

Although there are some "virtual command center" software programs available, much of what they provide is redundant to what companies with best-in-class programs should already have with a third-party crisis response and intelligence provider such as iJET, Control Risks Group, or Drum Cussac.

Define reporting and status process

Acknowledgement of the situation, along with regular updates approved by senior leadership should be communicated to the company and sometimes externally via approved channels, including, but not limited to, intranets and extranets.

Hopefully, before an incident takes place, because of the travel and/or asset risk management platforms that employers have in place, a quick list of potentially impacted individuals and facilities can be identified and be used as the basis for a communications list in addition to key stakeholders and executives. These traveler tracking or asset risk management platforms should have the ability to not only automatically forward relevant alerts provided by your intelligence provider, along with updates, but also provide a conduit for any manual messages from your command center or key stakeholders to be sent via e-mail and/or SMS text message. Having this capability is a must. If you do not have it, and are procrastinating about obtaining these features, as well as others, such as trip briefing disclosures, you are doing a disservice to your organization and its travelers and potentially setting yourself up for excess liability.

Conduct crisis "readiness" exercises

There are four primary areas of crisis readiness to base your exercises on:

1. Emergency Management—Generally day-to-day processes for incidents that can routinely be addressed via third-party support and defined protocols. This is often managed in concert between security (if applicable), human resources, and your third-party crisis response provider.
2. Crisis Communications—The process by which an alert or incident exceeds the company's threshold for acceptable risk and must be escalated and communicated to the general population and/or targeted individuals.
3. Crisis Management—How crisis response teams and command centers, along with third-party support, handles a variety of use cases from pandemics, civil unrest, war, terrorism, and natural disasters, to politically sensitive situations that could risk lives, investments, or the company's reputation.
4. Security Management—How a breach of confidential data via loss or theft of intellectual property is handled, as well as how to react to any lapse or breach in safety or service protocols for things like executive protection and physical security assets and services.

Continuous process improvement

Without a structured framework and plan toward travel risk management, and the means for measuring performance with metrics, continuous process improvement isn't possible. The industry standard for a common framework is the TRM3 (Travel Risk Management Maturity Model) as mentioned and discussed in previous chapters.

Even though each company's approach and protocols may be different in each key process area of the TRM3, if your program collectively addresses each of the KPA's, you should have the basis for measuring improvement year over year when benchmarked against other companies.

In consideration of the TRM3 key process areas, some examples to be considered or reviewed year over year with respect to continuous process improvement are as follows:

- Have your total number of emergency medical and security claims decreased?
- Has the total number of annual calls to your crisis hotline gone down, because of training, pretrip mitigation, or for some other reason?
- What is your average response time for accounting for any impacted travelers once a crisis communication has been distributed/received (e.g., from the moment that employers are notified of an incident, until they have tracked and accounted for all potentially impacted travelers)?
- How many incidents were supported by existing protocols and third-party support services, versus requiring escalation or exception support?

Getting the most out of your crisis hotline and response support provider

Assuming that the provider that an employer uses for crisis hotline and response support, is the same provider used for intelligence and traveler tracking and disclosure automation, there are still distinct differences between major providers in the market. Buyers should intimately understand the differences between providers, and consider the following:

- What is its core business and what does it outsource or partner with others for in its offering? (Is its focus medical services and evacuation memberships, intelligence, or physical security services?)
- What can your provider offer you to make up for gaps in your program? Does it have a global network of both medical and security service providers at its disposal? Are these resources its own or outsourced?
- If it can offer third-party products and services to you that fill gaps in your program, what benefits can you obtain by leveraging the provider's relationships with those other providers in terms of discounts or terms and conditions?
- If you have your own suppliers for different types of services (e.g., executive protection, medical support and evacuation), will your provider work with and support those suppliers, even if it has its own?
- If you are starting from nothing and working to create your first travel risk management program, will it help you to develop some of your protocols?

Kidnap and ransom, extraction and evacuation

6

Kidnap and ransom

Industry is expanding to remote parts of the world, and companies are sending travelers and expats to high-risk and lower-risk destinations with high-risk factors such as extreme poverty or religious extremism. However, the absence of travel or business in a high- or significant-risk location can no longer excuse an employer from having a plan in place, should a kidnapping occur. For instance, if you think that Islamic extremism is confined to a few countries in the Middle East, think again. While there are many other smaller groups, "Radical Islam" is organizationally composed of Hamas, ISIS, Al Qaeda, Taliban, Hezbollah, and Boko Haram, just to name a few, and their geographical reach is broad and far beyond just the Middle East, Africa, and Asia. According to CNN,[1] Hezbollah has been infiltrating Latin America for funding via money laundering, counterfeiting, piracy, and drug trafficking.

Let's assume that you're one of those people who attest, "We don't travel to high-risk destinations, so we don't need to worry about our people getting kidnapped," and then it happens. How did it happen? What do you do? Are you prepared? Do you have access to resources and services that can assist you with the careful process that you must now go through to try and ensure your traveler's safe return? What training and protection do you provide your travelers and expats (including their families) that reduce the possibility of abduction? These are all good questions to consider, but companies won't if they don't have a comprehensive plan that addresses, via policy, some of the potential times when and where someone may possibly be kidnapped because of afterhours behavior, location, acquaintances, or personal side trips.

The National Institute of Statistics reports that within the first 9 months of 2013, more than 105,000 kidnapping incidents took place, which numbers can be deceiving because of the large number of unreported incidents that occurred in places such as Mexico. According to the *Huffington Post*, more than 90 percent of abductions in Mexico go unreported.[2] Abductions can impact employees at work, home, or while traveling domestically or internationally.

---

[1] Arthur Brice, "Iran, Hezbollah Mine Latin America for Revenue, Recruits, Analysts Say," CNN, June 3, 2013, http://www.cnn.com/2013/06/03/world/americas/iran-latin-america/.

[2] E. Eduardo Castillo, "Mexico to Launch Anti-Kidnapping Campaign," *Huff Post Latino Voices*, January 29, 2014, http://www.huffingtonpost.com/2014/01/29/mexico-anti-kidnapping_n_4686456.html.

Kidnapping for ransom has become the largest source of terrorist financial support. According to the *New York Times*, because the income has proven so substantial as a global business, Al Qaeda in the Islamic Maghreb, in northern Africa; Al Qaeda in the Arabian Peninsula, in Yemen; and the Shabab, in Somalia, are coordinating their efforts and abiding by a common kidnapping protocol.[3]

According to the same article, citing a *New York Times* investigation, kidnapping Europeans has bankrolled Al Qaeda operations across the globe, citing that at least US$125 million has been paid since 2008, despite denials of paying ransoms by European governments. Of that amount, US$66 million was paid in 2013 alone. Additionally, the United States Treasury Department has cited collective ransom amounts for that same period at greater than US$165 million, made almost exclusively by European governments, paying the money through proxies as foreign or developmental aid.

The New York Times goes on to say that to minimize risk to their fighters, terrorist groups have outsourced the seizing of hostages to criminal groups who work on a commission, taking a reported 10 percent of the ransom.

Suffice it to say that while the numbers associated with terrorist-related groups are alarming, they don't include the massive number of kidnappings merely for financial, non–politically motivated gain.

For a more global view of the number of reported cases to the police, the United Nations Office on Drugs and Crime publishes annual kidnapping statistics that can be found online at https://www.unodc.org/unodc/data-and-analysis/statistics/crime.html.

According to Gary Noesner, former Chief, FBI Crisis Negotiation Unit, and author of "Stalling for Time: My Life as an FBI Hostage Negotiator," training and an individual's responsible behavior alone can drastically reduce the potential for abduction. It's part common sense and part training. For instance, traveling alone in a high-risk country, a traveler may be faced with choices such as spending a few extra dollars on ATM fees by using the ATM machine in the hotel versus down the street, or they may not like the hotel food versus venturing out alone at night. In both instances, do the right thing and use the hotel ATM and eat there. Don't take unnecessary chances. Knowledge is power and employers need to provide travelers with intelligence that educates them regarding what is risky and what is not. Mr. Noesner goes on to add that making the risk disclosures available to travelers to read at their discretion is no substitute for someone physically sitting down with them and briefing them prior to travel to a high-risk destination.

Mr. Noesner goes on to say that how someone *should* react to a situation could depend on where they are located at the time. For instance, a woman alone in the United States walking to her car in a parking lot that is faced with a van that pulls up and has an armed man demanding that she get in, should, in his opinion, never get into

[3] Rukmini Callimachi, "Paying Ransoms, Europe Bankrolls Qaeda Terror," *The New York Times*, July 29, 2014, http://www.nytimes.com/2014/07/30/world/africa/ransoming-citizens-europe-becomes-al-qaedas-patron.html?_r=0.

the van. Generally, where such a confrontation occurs, the goal is most likely sexual predation, with lethal consequences. Instead, she should yell, scream, call for help, and run away. The abductor is looking to convince her through threats to get in the vehicle, not to have a confrontation in the parking lot. He is not looking to commit homicide there and then. Conversely, if the same thing were to happen in Mexico City, the motivation is almost always money and she should avoid extreme resistance based on the likelihood that the kidnapper will turn violent. She should instead comply, with the knowledge that statistically she will be negotiated out by her family or employer. While this prospect is undesirable, it is her best path toward staying alive and eventually gaining her freedom.

In the situation where someone believes they are being followed, it is best that they are not timid and look away, but instead make direct eye contact and stare down whomever is following. Criminals do not want to be noticed or remembered by their victims. Awareness of their stalking behavior can serve to discourage their continuing interest in you. Being able to identify an assailant, taking away their anonymity and element of surprise, makes you an undesirable target.

Since 2008, Al-Qaeda has earned an estimated US$125 million dollars in ransom payments for hostages, with much of those funds from Europe. BBC News has produced a 60-second video, "Which countries pay ransoms for hostages? — In 60 seconds," on which countries pay ransoms for hostages, that is worth viewing at http://www.bbc.com/news/world-us-canada-33265309.

Types of kidnapping

1. Kidnap for ransom
2. Express—brief captivity, typically for the purpose of fast cash from ATM or bank withdrawals.
3. Virtual—threat of, or staged abduction, for quick extortion before discovery that the threat has been removed or abduction has not actually taken place. According to Vice.com,[4] in July 2014, Mexico City's Policy for the Prevention of Cybercrimes issued an alert about the number of extortions via Internet messaging applications, advising users to refrain from posting personal contact information and to keep their profiles private. Perpetrators can use personal information, photos, and actors portraying loved ones (screaming in the background) making phone calls or posting audio files, to lead victims into believing that they or their loved ones are in danger. Timing and details are essential for these crimes, who are searching for a quick payout before victims contact loved ones directly or the authorities. Journalists have been known to be targeted by criminals for exposing the criminals' activities, and have been met with violence.
4. Revenge—usually for political or religious purposes

---

[4] Priscila Mosqueda, "Mexican Drug Cartels Are Using Social Media Apps to Commit Virtual Kidnappings," Vice, September 17, 2014, http://www.vice.com/read/mexican-cartels-are-using-social-media-apps-to-commit-virtual-kidnappings-917.

How to Avoid Becoming a Virtual Kidnapping Victim

To avoid becoming a victim, look for these possible indicators:

- Callers go to great lengths to keep you on the phone, insisting you remain on the line.
- Calls do not come from the victim's phone.
- Callers try to prevent you from contacting the "kidnapped" victim.
- Multiple successive phone calls.
- Incoming calls made from an outside area code.
- Demands for ransom money to be paid via wire transfer, not in person; ransom demands may drop quickly.

If you receive a phone call from someone demanding a ransom for an alleged kidnap victim, the following course of action should be considered:

- Try to slow the situation down. Request to speak to the victim directly. Ask, "How do I know my loved one is okay?"
- Ask questions only the victim would know, such as the name of a pet. Avoid sharing information about you or your family.
- Listen carefully to the voice of the kidnapped victim if they speak.
- Attempt to call, text, or contact the victim via social media. Request that the victim call back from his or her cell phone.
- To buy time, repeat the caller's request and tell them you are writing down the demand, or tell the caller you need time to get things moving.
- Don't directly challenge or argue with the caller. Keep your voice low and steady.

Source: The FBI (Federal Bureau of Investigation), *Stories*, "Avoid Becoming a Victim of Virtual Kidnapping," November 4, 2014, http://www.fbi.gov/news/stories/2014/november/virtual-kidnapping/avoid-becoming-a-victim-of-virtual-kidnapping.

Commonsense kidnapping avoidance tips

1. In some countries, there have been increased reporting of driver abductions, with the intent to abduct their arriving passengers. Consequently, companies should establish a structured secure ground transportation program that includes:
    a. Policies and procedures for reserving and interacting with both security-based (utilization of personal security services for ground transportation) and non–security-based ground transportation providers.
        i. Where and when security-based ground transport can be used.
        ii. Suppliers to never post the company name on name placards at pickup locations (use branded name for the company's security program instead, as it is less obvious).
        iii. Supplier to never post the traveler's name on name placards at pickup locations (instead, both the driver and the passengers should have names and photos of each other available for identification of one another in a public place).
        iv. Consider the use of passwords for drivers that are unique to each trip.

2. Employ a buddy system when possible—travel in pairs or groups whenever reasonable or possible. This can often reduce the risk of abduction. People traveling alone are often the easiest targets for kidnappers.

3. Inform people of your schedule and travel plans—Always provide someone with a detailed itinerary for your trip, including meetings throughout the day if possible. This can narrow down the potential time and place of abduction if it takes place.

4. Maintain a low profile—avoid sharing personal information with people you do not know, including drivers, service people, and hotel staff. Especially when it comes to your nationality.

5. Always carry a mobile phone—additionally, if your company provides the program/ service, have your company's security services provider's mobile application installed on your phone with location-based services enabled in your settings. Often, these apps have a panic button on them that can transmit a "need help" message or dial a crisis hotline, whereby they can either lend assistance, or identify that you are in trouble and use GPS tracking from your phone to find you.

6. Does your mobile phone work where you are going?—consider your company's international phone network: Will providers support your plan in the travel area or do you require the use of satellite phones.

7. Limit or avoid alcohol intake—alcohol can impair your judgment and inhibit travelers from making good decisions or possibly defending themselves when necessary.

8. Avoid public transportation when alone—use a company-approved car or taxi service. Do not take random taxis if possible, and never hire an unmarked sedan service or "pirate driver" found at some unregulated, metropolitan airports.

9. Avoid repetitive patterns—try to dine at different places on different days of the week, take different routes to and from work, at different times during the day, and alternate hotels in cities that you visit frequently.

10. No branded clothing or merchandise—advertising that you work for a western company via branded scarves, shirts, or jackets is a welcome invitation for attention from criminals and terrorists.

11. If concerned that someone is following you via car, always circle roundabouts or neighborhoods twice to confirm suspicion. Once confirmed, proceed to a police station or populated location, while noting the make, model, and license plate details of the car, if possible.

12. Always ensure that any hired local help have been properly screened with background checks prior to employment.

Kidnapping crisis response—preparation and training

Establish a framework and process flow for a kidnap and ransom response plan, assigning key stakeholder responsibilities, often modeled after recommendations provided by your kidnap and ransom insurer. Companies will want to make sure that their actions are not only legally sound and in the best interest of the victim, but also in compliance with their insurance coverage as much as possible to ensure that coverage is not sacrificed or nullified by not following procedure.

Proof of life and kidnapping policies and procedures

Proof of life is a concept that means that a company has a "proof-of-life" policy and procedure with questions and answers on file for designated employees, so that in the circumstance where someone may have been abducted, asking very confidential "proof-of-life" questions that only the traveler will know the answers to, will help indicate to the employer that, indeed, their traveler has been abducted and that, for the moment, the traveler is still alive.

Suggestions for Developing a Proof-of-Life Questionnaire

by European Interagency Security Forum (http://www.eisf.eu)

Choose questions with answers that:

- Are available to a small group only.
- Are easy for the hostage to remember in times of stress.
- Recall happy times and lift the hostage's morale, for example:
  - What was the name of your childhood holiday home in California?
  - Where did you meet your husband/wife?
  - What is your brother's/sister's nickname?

Avoid questions on topics that a captor may find provocative, for example, questions about:

- Religion
- Politics
- A lifestyle that might be viewed as "decadent"
- A previous job, for example, with military connections

Avoid questions to which the answers are easily found on social networking sites, for example:

- Date of birth
- First school
- Name of boyfriend/girlfriend/pet

Having a proof-of-life policy and procedure in place can not only save time, but money as well for employers because cases have been reported in which captors cut off a body part from an abductee and sent it to the employer to get the employer's attention. The loss of a body part might not have happened had a proof-of-life procedure been in place. Such a loss to an abductee is likely to result in a lawsuit against the employer for damages caused by the employer's lack of preparedness.

Communication

A communication checklist should be on hand for a kidnapping situation that addresses the following:

1. Who writes all official communications regarding the incident, regardless of the recipient?
2. What are all the communication requirements for key stakeholders, such as authorities, insurance providers, professional security personnel (negotiators), company legal counsel, and senior management at headquarters?

3. Who approves all written communications?
4. What information does the company provide to the local and national government authorities and media, and when?
5. What information does the company provide to the family of the victim and when (as well as how frequently updates are provided)?
   a. Status of the situation
   b. Efforts to resolve the situation
   c. Resources available to the family during the crisis
      i. Will the victim still get paid during captivity?
      ii. What expenses does the company cover during the crisis?
         1. Recovery
         2. Repatriation of mortal remains (if applicable)
         3. Psychological counseling support
         4. Any other logistical, financial, emotional support
         5. What information is provided to coworkers and when?

Interaction with the victim's family

According to Gary Noesner, former Chief, FBI Crisis Negotiation Unit and author of "Stalling for Time: My Life as an FBI Hostage Negotiator," in the event of a kidnapping, a victim's family needs assistance in two primary areas: (1) operational information and (2) emotional support. Too often, companies will send out a Human Resources (HR) representative to help the family with monetary support, relocation assistance, benefit claims, and other logistical needs, but not to inform them of operational updates and initiatives. He adds, the HR or EAP (Employee Assistance Program) model fails to address the family's strong need for operational information, as they will want to know what the organization is doing to secure the freedom of their loved one. Mr. Noesner goes on to add that sending an HR/EAP representative alone is as problematic as supporting the family with operational updates but without emotional support.

A blended model of both is recommended, typically including someone who can assist with logistics and counseling support, as well as a security executive who is trained in how to communicate with families about how the investigation and negotiations are proceeding, what the company's plans are, and to provide regular updates in a timely fashion. Families must feel as engaged as possible while the professionals involved do their jobs, and also feel strongly that the company is doing everything possible to get their loved one released without further incident.

Traditional and social media roles relative to kidnapping

Traditional media, such as television, newspapers and radio, often remain silent with a story regarding a kidnapping, allowing the parties involved to try and resolve the incident as quickly as possible, with the rare exception of a high profile case, often with political or journalistic ties. However, with social media, news of the capture by assailants can be broadcast across social media via Facebook and Twitter accounts within moments of the kidnapping.

Along with HR/EAP and security resources assigned to the families of kidnapping victims, companies should provide a public relations resource to manage traditional and social media relations and to coach the family, should the family be approached by reporters on their own.

Social media training for travelers, expats, and family members should include, but not be limited to, the following:

1. An understanding of how kidnappers use social media to target their victims, obtain current pictures of travelers or expats and their loved ones, as well as personal information, including contact details and other information that can be used against the victims, both before and after a kidnapping.
2. How to set social media security settings so as prohibit access to user content except to an authorized viewer or contact. In those circumstances where the media in question doesn't allow that, teach them what not to post.
3. The dangers of using social media to communicate with kidnappers, sharing personal information regarding the situation, or reacting to false or even accurate information provided by someone on social media.

Hostage survival training

If a company operates industrial manufacturing involving hazardous chemicals, it is highly likely that it would provide training to employees and others who are or will be exposed to or at-risk for a spill or exposure. Similarly, it is important for those travelers and expats going to an area deemed as a high risk for kidnapping to receive survival training to help them stay alive should they get kidnapped.

Hostage survival training may vary from one program to another, but are often largely based upon the psychological methods and tools used to help victims get through the various phases of their experience with as much of their sanity as possible. Each of the following phases of a kidnapping, and how the victim should handle each phase, should be included in the training:

1. The assault and capture phase
2. Transport phase
3. Captivity phase
4. Release and reintegration phase

Insurance

Kidnap and ransom (K&R) insurance is a policy intended to reduce the financial impact of a K&R incident on your company. Some of the items that may be covered by such policies are:

- Negotiator fees
- Investigators
- Attorneys
- Public relations professionals
- Forensic analysts
- Security guards and consultants
- Reward money
- Medical costs not covered by standard insurance
- Psychiatric treatment
- Loss of employee income

- Childcare
- Special investigations equipment
- Rehabilitation
- Funeral expenses

K&R insurance is often referred to as "The best insurance no one should ever know that you have." The reason for saying this is that if companies have this, it is in their best interest to keep this a well-guarded secret and not to tell anyone because such knowledge could draw more attraction to travelers and put them at greater risk if criminal elements knew that the coverage existed.

However, the potential for release of a hostage today relies heavily on the nationality of the victim, and whether or not the victim's employer has K&R insurance. According to the United Kingdom's (UK) *The Guardian*,[5] using data provided by Filon's AKE Group (a British security firm that collects kidnap and ransom information from around the world), with increased numbers foreign oil and mining contractors and aid workers in the Sahel region, the average westerner can be ransomed for US$3.75 million. The article goes on to state that in 2014, at least 75 percent of Fortune 500 companies hold K&R insurance policies. While there are no statistics on K&R policy sales, since 9/11 and during the period after the Arab Spring, policy sales reportedly increased dramatically. According to *The Guardian*, in 2009, a company buying kidnap insurance in Nigeria would pay approximately UK£6,000 per year (US$10,000) for UK£3 million (US $5 million) in coverage, according to Willis Special Contingency Risks, a unit of insurance broker Willis Group Holdings in London. However, now that same coverage sells for approximately UK£60,000 (US$100,000).

Reports state that ransoms for American journalist James Foley and two other Americans were in excess of US$132 million after their kidnapping in 2012. However, the U.S. government's policy against paying ransoms and the facts that they were Americans and Mr. Foley didn't have insurance, made the situation even more difficult.

Reaction

What should you do when a kidnapping occurs?

Just as with your crisis response and business continuity plan, and just as you would have a pandemic plan, you should also have a supplemental kidnap and ransom plan, calling to action your crisis command center (with the ultimate decision maker usually being the CEO in the case of a kidnapping), a primary coordinator (often a corporate security director or risk executive) to collect and organize information from stakeholders, and your legal counsel.

When an incident occurs:

1. Collect as much information about the circumstances surrounding the abduction, the victim's medical condition at the time, if known, and full disclosure of any communications from the assailants.

---

[5] Derek Kravitz and Colm O'Molloy, "The Murky World of Hostage Negotiations: Is the Price Ever Right?," *The Guardian*, August 25, 2014, http://www.theguardian.com/world/2014/aug/25/murky-world-hostage-negotiations-price-ever-right-insurance.

2. Advise person reporting the abduction and any non–crisis management team individuals that are aware of the incident, to not speak to or inform the press or law enforcement authorities until the crisis management team leadership decides to do so. Depending upon the location, advisors may suggest not including local law enforcement authorities because of corruption and the potential to further complicate matters.

3. Any staff members who receive communications from assailants should record any conversations when possible. Such staff members should listen only and not try to negotiate. Any and all information regarding the situation should be provided to the crisis management team, which is where all decisions regarding the incident should be made.

4. Assemble an immediate meeting of the crisis management team, and trigger your kidnap and ransom response plan, which will typically include quick decisions such as:

    a. Assessment of the situation

        i. Understand what the assailants want and when

            1. Money?

            2. Political action?

            3. Change in business practice?

            4. Religious reasons?

        ii. Is it clear where the hostage is being held (general geographic location)?

        iii. What resources do you have available in the area that might prove helpful?

            1. Local politicians and law enforcement (if helpful and relevant)

            2. Tribal and religious leadership ties

            3. Investigation and extraction services

    b. Who to contact and when (law enforcement, consultants, insurance providers, family members, etc.)?

    c. Who should communicate with the assailants, how and when?

Time is of the essence in a K&R situation and can be the difference between life and death, so being able to execute a plan and work toward a solution without delay is critical. Although K&R insurance responders provide policyholders with options and scenarios, they won't make decisions for the policyholders. This is another reason why having a plan and a designated team for these situations are important parts of your business continuity and crisis response plans.

Government support or lack thereof for ransoms

Even though it is not possible to list the positions of every country around the world on the topic of governmental support of kidnapping hostage recovery and ransoms, the following examples will provide some good insight into how complex an issue this is legally, and stress the importance of corporations having a plan in place that includes understanding such legal and political positions that may impact their decisions and processes.

The United States

As of June 2015, the Obama Administration changed policies to no longer criminally prosecute families who pay ransoms to get loved ones back from groups such as ISIS. However, the following excerpts from U.S. policy are what was on record prior to this decision and should be considered when a company or employer decides to pay such ransoms, with due diligence taken into account as to whether or not they would get the same relief from prosecution as the family members of kidnapping victims.

According to the U.S. Department of State,[6] prior to June 2015, the U.S. government believed that paying ransom or making other concessions to terrorists in exchange for the release of hostages increased the danger that others will be taken. Its policy, therefore, rejected all demands for ransom, prison exchanges, and deals with terrorists in exchange for the release of hostages. At the same time, it made every effort, including contact with representatives of the captors, to obtain the release of the hostages.

The United States strongly urged American companies and private citizens not to pay ransom. It believes that good security practices, relatively modest security expenditures, and continual close cooperation with embassy and local authorities can lower the risk to Americans living in high-threat environments.

Although the U.S. government was concerned for the welfare of its citizens, it could not support requests for host governments to violate their own laws or abdicate their normal law enforcement responsibilities. On the other hand, if the employing organization or company worked closely with local authorities and followed U.S. policy, U.S. Foreign Service posts could actively pursue efforts to bring the incident to a safe conclusion. This included providing reasonable administrative services and, if desired by the local authorities and the American organization, full participation in strategy sessions. Requests for U.S. government technical assistance or expertise was considered on a case-by-case basis. The full extent of the U.S. government participation had to await an analysis of each specific set of circumstances.

If a U.S. private organization or company sought release of hostages by paying ransom or pressuring the host government for political concessions, U.S. Foreign Service posts would limit their participation to basic administrative services, such as facilitating contacts with host government officials. The host government and the U.S. private organization or citizen must understand that if they followed a hostage resolution path different from that of U.S. government policy, they do so without its approval or cooperation. The U.S. government cannot participate in developing and implementing a ransom strategy. However, U.S. Foreign Service posts can maintain a discreet contact with the parties to keep abreast of developments.

Under current U.S. law 18 USC 1203 (Act for the Prevention and Punishment of the Crime of Hostage-Taking, enacted October 1984 in implementation of the UN Convention on Hostage-Taking), seizure of a U.S. national as a hostage anywhere in the world is a crime, as is any hostage-taking action in which the U.S. government is a target or the hostage-taker is a U.S. national. Such acts, therefore, are subject to investigation by the Federal Bureau of Investigation and to prosecution by U.S. authorities. Actions by private persons or entities that have the effect of aiding and abetting the hostage-taking, concealing knowledge of it from the authorities, or obstructing its investigation, may themselves be in violation of U.S. law.

---

[6] U.S. Department of State, Bureau of Public Affairs, "Fact Sheet: International Terrorism—American Hostages ," October 17, 1995, http://dosfan.lib.uic.edu/ERC/arms/cterror_briefing/951017cterror.html.

According to the *U.S. Department of State Foreign Affairs Manual Volume 7 Consular Affairs* (7 FAM 1820 HOSTAGE TAKING AND KIDNAPPINGS[7]–CT:CON-526; 08-11-2014–OFFICE OF ORIGIN: CA/OCS/L), references for emergency planning guidance, definitions for "Hostage taking" as defined by the U.S. government, treaties outlining consular authorities in providing assistance in hostage taking or kidnapping situations, the role of consular officers, the Federal Bureau of Investigations (FBI), Department of Justice and other government agencies in said situations, and many applicable references to legislation, are provided at great length.

U.S. law on kidnapping ransom offenses, with definitions and potential sentence guidelines can be found on the United States Sentencing Commission website at the link provided below.[8]

United Kingdom

UK law provides that the payment of a ransom is not an offense as such, although the government itself will not make or facilitate a ransom payment, and will always counsel others against any such substantive concessions to hostage takers. However, Section 15(3) of the Terrorism Act 2000 makes it an offense for a person to provide money or other property if the person knows or has reasonable cause to suspect that it will or may be used for the purposes of terrorism.[9]

Australia

They key policy shaping the Australian government's response to the kidnapping of Australian citizens overseas is that the government does not pay ransoms.[10]

According to smartraveller.gov.au, the Australian government's ability to provide consular assistance to Australian citizens may be severely limited in locations where the Australian government recommends against all travel and in places where the security situation is particularly dangerous or access is limited.

Should an Australian be kidnapped, the Australian government will work closely with the government of the country in which the kidnapping has taken place, as well as other governments, to ensure that all appropriate action to resolve the situation is pursued actively. The Australian government will provide information to families on what they can expect and provide them with clear and up-to-date information on developments in the case to help them make informed decisions.

The Australian government does not make payments or concessions to kidnappers. The Australian government considers that paying a ransom increases the risk of further kidnappings, including of other Australians. Ransom payments to kidnappers,

---

[7] http://www.state.gov/documents/organization/86829.pdf

[8] United States Sentencing Commission, "2011 Federal Sentencing Guidelines Manual, Chapter Two–Offense Conduct," http://www.ussc.gov/guidelines-manual/2011/2011-2a41

[9] "Money Laundering and the Financing of Terrorism – European Union Committee Contents, Annex A," http://www.publications.parliament.uk/pa/ld200809/ldselect/ldeucom/132/9031112.htm.

[10] Parliament of Australia, "Part II: Response to Kidnapping Incidents, Chapter 4: Australia's No-Ransom Policy," http://www.aph.gov.au/Parliamentary_Business/Committees/Senate/Foreign_Affairs_Defence_and_Trade/Completed%20inquiries/2010-13/kidnapransom/report/c04

many of whom are associated with proscribed terrorist groups, are also known to have funded subsequent terrorist attacks.

Extraction

"Extraction" is defined as "the act or process of getting something by pulling it out, forcing it out, etc." Extraction should not be confused with the term "evacuation," which is discussed separately. For the purposes of travel risk management, we briefly address these terms in the following contexts:

* Kidnapping victim extraction
* Nonmedical, security extractions

In an interview, Tom Winn, lecturer and program director of the Master of Security Management for Executives, University of Houston–Downtown, stated the following:

> There is a common misconception that when someone is kidnapped a team of government special operators will be en route in a few days to extract the hostage. The reality is quite the opposite. Some movies and more recent press coverage have shown a somewhat more realistic view, but it is still often a "Hollywood" version. In truth, just like decisions to protect travelers should be based on risk, the decisions concerning kidnap response should also be based on risk. There are several consulting organizations who have a well-deserved quality reputation in kidnap response consulting. Their methods have been proven over years of successfully working through kidnap crises around the world. The responders are well trained and consult with the victim's family and/or company on the entire crisis response.

"Sending a team" into a location to extract a hostage is a last resort and rarely a legitimate option. Such an operation is fraught with tremendous risk to the hostage, the organization, the families, and the team of operators. There are also many real and potential legal issues that must be discussed. Some of the risks that should be taken in to consideration and questions that need to be answered before such an operation are discussed below.

Risks to the operation

Who is actually going to perform this operation (i.e., who will make up the team)? There are some private military contracting firms (PMCs) who claim they can do this. They are generally made-up of, or at least led by, a former special operations team member: U.S. Military Special Ops, British SAS, German GSG 9, etc. These individuals are very capable and the teams of which they were members are some of the best in the world at responding to this type of crisis. However, what many people do not think of is those special operations teams have all the support of their government's intelligence, military, diplomatic, and legal apparatus. What does the private team have to support it? How will the team move in **and out** of a country? How will the hostage be located and will that hostage be there when the operation starts? How do you verify the experience and capability of the contractor you have selected? Can that firm really do what they say they can do?

Legal risks

Is the team legally allowed to perform an operation of this type in the location? Most countries have very strict gun laws, especially for foreign visitors. Many nations also have very strict laws about who is allowed to be armed security and who is not. Furthermore, the locals may have a responsibility to report such activity to local authorities. Will there be a legal or punishment response to any local employees or nationals who have sponsored your expatriate or traveling employees? The home country may have laws that impact such a response, as well.

It is almost a guarantee someone will be hurt or killed. What are those legal ramifications? Other countries' legal systems are not the same as that of the United States. Trials are almost never in favor of the accused or even fair. Punishments are often considered "cruel and unusual" by U.S. standards. Prisons in many parts of the world are unimaginably harsh.

Risks to the organization

Does your organization hold to any of the U.N. humanitarian doctrines of social responsibility or standards of behavior for security companies? Violating one of the doctrines can, at minimum, result in the loss of a significant contract or multiple contracts with governments and/or aid agencies. Can your organization stand the legal, commercial and social pressure if an innocent person is hurt or killed accidentally?

Extracting a hostage is fraught with risk and even greater loss of control of a bad situation. It is best to leave that option to the expert government teams and their extensive resources.

Evacuation

An evacuation is defined more by the actions or assisted actions of leaving a dangerous place, whether that be for security or medical reasons, and taking evacuees to a safer location to either shelter in place or receive superior medical treatment or care, whatever the case may be.

Medical evacuation plans on the consumer and commercial markets are a big business globally and vary widely in terms of cost and coverage. In general, consumer-based travel insurance are often not ideal options for business travelers and expats. Consumer-based policies often are based on a per-trip premium and include coverage that is often unnecessary or duplicate in nature because of existing similar coverage via corporate card programs or business travel accident (BTA) plans, such as trip cancellation coverage and lost luggage insurance, all of which are ultimately part of the premium you pay for, when all you may want to focus on is emergency medical treatment and evacuation or security coverage.

Key considerations when evaluating consumer-based travel insurance:

1. Are you paying for coverage that you don't need (e.g., trip cancellation insurance, lost luggage, etc.)?

2. Does the plan offer guaranteed payment for medical treatment overseas, or only reimburse you upon your return (subject to review and approval)?
3. Is the plan strictly for medical evacuation? Does the plan cover any medical diagnosis or treatment?
4. Who makes the determination to evacuate?

Nonmedical security evacuations are typically used when a conventional evacuation via commercial or private transportation is not possible without the intervention and logistical support of a third-party security services firm. In other words, you must physically be removed from a location. Some examples of when a nonmedical security extraction may be needed are:

1. Removal from an area where airports have been closed for various reasons, and otherwise safe ground transportation has become dangerous.
2. Removal from a location where infrastructure has broken down because of a natural disaster, civil unrest, or some other reason.

Key differences in corporate medical and security plans and insurance

Access to medical network

Many companies that already have some type of policy to cover payment of medical emergencies and/or evacuations don't need additional insurance, but do need access to a global provider that has comprehensive resources, facilities, personnel, and access to medical transport. In those circumstances, the company pays an annual "access to medical network" fee, which is sometimes viewed as a membership. This, in effect, outsources support for medical treatment, support, evacuation, and, depending upon the provider, security support, evacuation, extraction, and related services. The company uses third-party insurance, unless self-insured, to pay for fees and actual expenses assessed during the year, in addition to the access fee.

Access to medical network plus insurance

This is in essence the same thing as "access to network," but with a supplemental insurance product to indemnify the company from costs within coverage for expenses incurred during the coverage year, offered by the network provider.

Access to security network

For those companies whose medical providers do not offer security services, extraction and evacuation coverage programs are offered that allow the company to pay an annual fee for access to a network of global security service providers. Typically, case fees and expenses incurred are additional costs. Like with access to medical networks, these expenses can be paid directly by the corporation or via third-party insurance policies for the specific services required.

Access to security network plus insurance

The same as access to a security network, but with a supplemental insurance product to indemnify the company from costs within coverage for expenses incurred during the coverage year, offered by the network provider.

Kidnap and ransom coverage

This is a policy specifically written for those circumstances where a traveler has been abducted. The kinds of expenses typically covered by these policies are discussed earlier in this chapter.

Hotel safety

7

This chapter looks at multiple aspects of hotel safety, from the perspectives of the buyer (employer), the traveler, and the hotel. Each perspective sheds unique light upon aspects of hotel safety and security that form the basis for a higher level of awareness and precaution when it comes to the hospitality industry.

Employer perspective

When staying at a hotel in the course of doing business, the hotel is in effect an extension of the workplace. Employers need to bear this in mind when evaluating preferred hotel suppliers or specific locations in conjunction with their preferred corporate hotel programs or during the site-selection process for meetings and events. Employers can't ensure that 100 percent of hotels used for employee travel meet all of their standards; thus the importance of adherence to a mandated corporate travel program that includes the use of preferred hotels, which hopefully meet an acceptable amount of the company's safety standards. One of the biggest challenges for employers in this area is consistency among providers. Many global hotel chains sell franchises to different owner-operators, and not all of them require the same safety resources and standards at all locations worldwide; therefore, just because you have become accustomed to a brand providing what you need in high-volume locations, doesn't mean that you can be certain that all of their hotels will meet the same standards.

Throughout the hospitality industry, the emergence of industry assessments is starting to gain some traction. For instance, OSAC (Overseas Security Advisory Council, a division of the U.S. Department of State), the Hotel Security Working Group, and the American Hotel & Lodging Association have developed and made available an online "Hotel Security and Safety Assessment Form," which can be found online at http://www.ahla.com/uploadedFiles/AHLAOSACHotelAssessmentForm2014.pdf.

In Part One of the assessment, covering background and property information, the assessment authors acknowledge that assessing hotels can be challenging because there are no widespread industry standards across hotels as a result of variable brand operational standards. Only general best practices are accepted, making an assessment subjective in the absence of a defined standard, yet more appropriate than an audit or inspection.

For instance, the following list includes what employers may want to consider when selecting hotel partners via a request for proposal (RFP) or otherwise for transient travel or meetings and event-based usage:

1. Emergency Response Procedures and Support
   a. Onsite security
      i. Where and when available?

      ii. In house versus contractors?

      iii. Applicable training?

  b. Police support

      i. How close and how accessible?

  c. Medical support

      i. How close and how accessible?

      ii. Onsite medical support?

  d. Evacuation or shelter-in-place plans

      i. Fire

      ii. Earthquakes

      iii. Active shooters

      iv. Bomb threats

      v. Civil unrest

2. Physical Security

  a. Fire-rated guest-room doors

  b. High-standard electronic room locks and secondary lock devices

  c. One-way viewing devices in hotel room doors

  d. CCTV (closed circuit television)

  e. Parking area security and lighting

3. Access Control

  a. Centralized control of locking systems

  b. Adequate distance control for street car access to lobby or hotel entrance doors (blast radius)

4. Guest Personal Information and Data

  a. Guest records

  b. Wi-Fi networks

5. Fire Safety

  a. Emergency response

  b. Fire suppression systems (access to extinguishers and sprinkler systems)

6. Preemployment Background Investigations

7. Continuous Process Improvement

  a. Metrics for measurement?

  b. How often?

8. Handicap Accessible

  a. Rooms

  b. Bathroom facilities

  c. Car parking

  d. Meeting rooms

  e. Emergency exits and resources

Hotelier's perspective

The hotel's duty of care to the safety of guests, in consideration of "reasonable best efforts," is to protect them from harm from reasonably foreseeable risks. This is a continuous legal duty, which if breached, gives a cause of action for negligence. The nature of the relationship between the hotelier (or innkeeper) and guest implies that the hotel represents the hotel premises were in reasonably safe condition.[1] However, hotels aren't necessarily the insurer of a guest's safety. In the U.S., their common-law liability may only apply to defects or conditions that are in the nature of hidden dangers, traps, snares, pitfalls, etc., which are not readily observable.[2] The hotel's duty is fulfilled when reasonable care is taken to prevent the invitee's exposure to dangers, which are more or less hidden and not obvious.[3]

In addition to consideration for safety and security requirements required by law or by buyers, understanding the property's challenges can help hotels to develop a pragmatic approach to addressing their duty of care. For instance, hotels should understand and monitor the levels of crime in the immediate area, potentially up to a few miles. Understanding the types of crime, such as vehicle vandalism and theft, can help with budgeting and planning for your program. This kind of data, in addition to your history of incidents onsite, can have a direct impact on budgeting. However, determinations will need to be made as to what level of security investment is made at each hotel for both personnel and technology, because it may not be enough to have video monitoring in place if the hotel doesn't have the staff to watch it.

It's important for hotels to also put a concerted effort into mitigating risks that may come from onsite staff and perhaps contractors. Criminal background checks, health and safety training, and security training is a growing requirement of many corporate buyers in light of reported incidents either assisted or even perpetrated by hotel staff. Travelers have an inherent trust in hotel staff to keep them safe; but once an improperly screened employee breaks that trust, perhaps because of a criminal background, the line of liability can usually be directly drawn to the hotel for not weeding out such applicants in the first place. News like that doesn't have to hit the press, because it spreads quickly through the corporate travel community. Making safety a deeply embedded part of a hotel company's culture, by doing things like offering rewards for reporting hazards, can promote continuous process improvement and make hotels more safe overall.

Training hotel employees

The following types of training, which contribute to a guest's actual or sense of safety, are some of those that hotels should consider for their employees:

1. Cultural sensitivity training—It's easy to discount this if you are a small hotel in a rural area, but in hospitality, you can rarely foresee with certainty who will show up and be staying with you. Having useful training in this area can often put guests from diverse cultures at ease and avoid incidents whereby a guest may become unruly if staff were unknowingly insensitive, thus putting others in harm's way.
2. Sexual harassment training—Understanding how to be mindful of conversations, behavior, and innuendo, and keeping them in a nonpersonal, business-like context is critical. Most people chat at check in, or the hotel bar, or while using other facilities with hotel staff, and

---

[1] *Rabon v. Inn of Lake City*, 693 So. 2d 1126, 1130 (Fla. Dist. Ct. App. 1st Dist. 1997).
[2] *Brown v. Alliance Real Estate Group*, 1999 OK 7 (Okla. 1999).
[3] http://caselaw.findlaw.com/ok-supreme-court/1387210.html

without proper training, the potential exists for even an innocent comment being taken out of context.

3. Minority (including LGBT [lesbian, gay, bisexual, and transgender]) sensitivity training.

4. Fire safety training—including evacuation plans and responsibilities.

5. Emergency medical training—While many hotels don't want to take on the responsibility of emergency medical assistance versus simply calling for local emergency support, simple measures like the Heimlich maneuver or cardiopulmonary resuscitation (CPR) training can save someone's life. This can be a controversial topic because of the amount of liability simply from getting involved, but with increased competition in many places and more critical buyers, such training, and even onsite medical equipment such as defibrillators, are being discussed.

6. Personal data privacy training—Ensuring that staff do not expose personally identifiable information to anyone carelessly or needlessly, on computer screens, in writing, or verbally. Such information must be safeguarded at all times and handled with extreme care and caution, exclusively in systems that have documented security and privacy standards.

7. Personal hygiene training.

8. Food handling and preparation training.

9. Security protocols and procedures for a variety of incidents, including effective recording and documentation.

In the Middle East and Africa, companies should consider hotels that have conducted in-depth analysis on the potential access by vehicle- or person-borne explosive devices, and the amount of damage that the structure could sustain in the event of such an explosion.

A recent trend has shown many hotels and hotel groups to lean toward the use of third-party assessments of hotel safety, such as the Safehotels Alliance AB.

Safehotels Alliance AB is a Swedish company founded in February 2001 and launched in October 2002. The company was formed after years of extensive research of the travel trade in general and the hotel trade in particular. Safehotels Alliance AB is formed and supported by experienced and highly respected individuals and organizations in the international travel and security industry. The company was created because safety is a primary concern in today's international hotel and meeting industry. According to several sources, however, there are not enough initiatives taken to meet the growing concerns. Safehotels' solution is to provide an objective third-party evaluation of business hotel and conference venues' security standards worldwide, covering all-important aspects of hotel and conference security. It currently provides assessments in 33 countries.

The Global Hotel Security Standard[4]

Safehotels has created, "The Global Hotel Security Standard", to meet the need for measuring hotel and conference security and matching the business travelers demand for it. Safehotels makes this matching extremely easy and cost efficient, and speeds up the communication process between the parties, the hotel and/or conference venue, and the buyer of hotel nights.

[4] The Global Hotel Security Standard is copyrighted and trademarked by Safehotels Alliance AB.

Work with the venues

The venues are audited individually. If the establishment meets the demands asked for in the Global Hotel Security Standard, it will become a part of the Safehotels network, stating that the venue has a well-managed security operation. The standard sets a high security level, which allows for local solutions. The Global Hotel Security Standard does not impose an exact way on how to solve safety and security issues, but do measure the quality of the solutions.

The venues provide good security management

Being a member of Safehotels Alliance AB gives no guarantees that an accident cannot happen; however, the risk of an unfortunate event has been greatly reduced by the venue's active participation in the Safehotels program. In addition, participants are prepared to handle an incident.

In an interview conducted with Safehotels Alliance AB CEO, Mr. Hans Kanold, we asked him the following questions about the organization, its process, and the importance of safety standards for hotels.

1. Tell me about the metrics and processes behind how you assess a hotel, and does it vary by request or customer type?
   We assess the hotels according to our "Global Hotel Security Standard." This standard was originally put together in 2001-02 in cooperation with travel managers, hoteliers, and security experts from a global perspective. The foundation thought and view is simply that "All guests should be assured that they hotel stay at has sufficient security standards," for example, as with equipment (i.e., fire alarms, locks, door closers, luggage rooms, burglar alarms, CCTVs), as well as a systemized way to maintain the equipment. Training, awareness, and documentation is also part of these standards. The Global Hotel Security Standard is reviewed and updated on a regular basis, and it contains approximately 220 checkpoints. The hotels are graded and awarded the appropriate certificate according to a specific scoring system. All hotels are looked at the same way, and that gives strengths to the certificate since it makes it easier to communicate to travel managers and tour operators. The reassurance that a third-party assessment is conducted yearly, checking the same items globally, gives confidence from a global approach. The guest or buyer decides if the level of certificate matches the demand at a certain destination the hotels carry. It is always better to stay in a certified hotel, knowing the hotel has been checked by a third-party assessor, than to stay in a hotel that doesn't have any proof of their security work, or only refers to their own system, which some hotels actually have. The hotels that are certified work proactively to avoid incident, and are also better prepared to react when an incident occurs. The challenge is the balance between security and comfort, as always.
2. Who are your primary or target customers, and do you see that evolving?
   We originally targeted working with mainly four- to five-star business hotels in largely known destinations when we started this 13 years ago, but the trend is that we do everything from resorts to also three-star hotels now, since the priority and demand from the travel industry has increased for these kinds of additional establishments. Our clients range from the small individual hotel to large international hotel groups like Rezidor Hotels, which has approximately 1500 hotels worldwide.

3. What are the different types of certifications that you offer? How are they different?

"There are three different certification levels: Safehotels Certificate, Safehotels Premium Certificate, and the Safehotels Executive Certificate. They differ in terms of equipment, density of security staff, and level of first aid preparedness, for example. The hotel is awarded from Safehotels to Safehotels Executive Certificate. The entry level is sufficient in all markets as a good base for security work; meanwhile the premium and executive certificates could be in more demand in certain, higher-risk destinations."

4. Are you seeing a rise in corporate buyer clients that want specific hotel inspections/ assessments?

"We see an increased demand from corporate buyers, absolutely. Our research in the field started 15 years ago, and the demand is rapidly rising. However, there is a clear demand for experienced consultants who do the audit on location. The "Fill in the blanks" self audit assessments do not fulfill their mission anymore. The demand is there for assessments for all events, but even more so for special events in particular with large numbers of participants in international locations. The destination and threat levels dictate the demand as well."

5. What are the demographics for those clients?

"Actually the demand for security standards in hotels is global these days. We don't just talk about terrorist attacks; we want the hotels to be able to handle also heart attacks as well, for example. Looking from a global perspective, that is more frequent and therefore just as important for the hotel to provide proper first aid in case of an emergency."

6. How long does it typically take from start to finish to conduct an assessment of one hotel?

"It varies. We ask the hotel to prepare, which we do as well, but then it depends on the size and type of operation. Normally 1 to 3 days on location."

7. Do your existing customers typically require recurrent assessments? If so, how often?

"To keep the certificate, the hotel has to go through a yearly visit from an accredited Safehotels Certification Consultant. It is a requirement. The hotel welcomes our assessment since it is like doing a yearly "Doctor's check-up." We help the hotels, and we remind them of possible spots where they need to improve, and we support the management and the hotel's security director. We have observed a healthy competition between certified hotels to increase their score year over year, which is good for everyone!"

8. Do you see the meeting and incentive industry starting to use safety or risk assessments more?

"Yes, the trend is clear, in particular because meetings are potential targets for terrorist attacks, but are also an occasion when incidents happen more frequently. The increase in international presence in meetings also brings specific challenges for all involved to be prepared for incidents large or small.

Some private intelligence and security firms provide standardized assessment and risk rating analysis for their programs on an ad hoc basis, but the Safehotels Alliance is the first of which to try and promote a global standard in the market place."

Rezidor Hotel Group Safety and Security "Always Care" Program Case Study

Similarly to how companies with traveling employees must adopt TRM as a discipline to be managed and improved upon, so must hotels adopt increasing standards for safety and security that covers their employees and guests across all geographies, cultures, and language barriers.

The Rezidor Hotel Group, based in Brussels, Belgium, has more than 430 hotels in 69 countries in Europe, the Middle East, and Africa, with 35,000 employees. Paul Moxness, Rezidor's Vice President of Safety and Security, is a leading pioneer and voice in the hospitality industry for driving change and improvement throughout the industry with regard to safety and security standards. Mr. Moxness is a founding member of OSAC's Hotel Security Working group and is also active in ASIS International, International Security Management Association, CSO Roundtable, and the European Institute for Corporate Security Management.

In 2013, Mr. Moxness had provided his leadership with a validated business case for why ongoing investment in global safety and security program standards across their network, provided a positive return on investment. He did this by showing a direct increase in corporate customer market share as a result of standards that he implemented across specific hotels in his group that specific, large, corporate customers required, which other hotel groups struggled to provide.

Always care–hotel questionnaire[5]
Protection of guest identity
- Objective: Guests will enjoy privacy while hotels will follow local legislation regarding record keeping.
- Risk Evaluation:
  - Do you know your local legal requirement for gathering guest information?
  - Do you work with IT and accounting to determine where and how any record containing personal guest information is stored and secured?
  - Is your staff trained in guest identity protection?

Control of guest identity
- Objective: Guest identity will be checked when necessary to meet legal or operational requirements.
- Risk Evaluation:
  - Do your check-in procedures ensure that guest name on credit cards and registration cards match the name in the reservation system?
  - Do you have procedures for checking identity if a guest asks for a new room key?
  - Do your procedures ensure that a new room key locks out previously issued room keys?

[5] Each organization must evaluate based upon its own business and individual circumstances what is considered best practices for what is to be included in a safety questionnaire. Rezidor makes no representation of whether this form is applicable or best practices for any other organization. This questionnaire version is dated 2014, and is subject to change.

Doors and locks
- Objective: Guests will enjoy privacy and not suffer any loss during their stay.
- Risk Evaluation:
  - Do you have electronic locks that keep an audit trail of events?
  - Are your guest room doors self-closing and self-locking?
  - Are your keys marked with hotel name and room number?

Procedures for accessing guest rooms
- Objective: Guests will enjoy privacy during their stay.
- Risk Evaluation:
  - Is staff trained in a specific procedure that makes them knock and identify themselves before entering a guest room?
  - Do staff ask for permission before entering a room when the guest is present?
  - Do you have a procedure for checking rooms that have a *Do Not Disturb* sign or that refuse service for a prolonged period?

Evacuation information
- Objective: Emergency procedure information will be prominently placed in all guest rooms.
- Risk Evaluation:
  - Are emergency and evacuation information cards placed on or near the inside of every guest room door?
  - Does emergency and evacuation information include a floor plan with pictograms showing exit routes, alarm buttons, and emergency equipment?
  - Is liability limitation information included on the evacuation information card?

Cloakrooms and locker rooms
- Objective: No guest will suffer any loss during their stay.
- Risk Evaluation:
  - Are cloakrooms monitored by employees or CCTV?
  - Are signs prominently placed to inform users that the hotel does not accept liability for loss?
  - Is liability limitation information included on receipts given to guests?

Parking facilities
- Objective: No guest will suffer any loss during their stay.
- Risk Evaluation:
  - Are parking facilities monitored by employees or CCTV?
  - Does the car park have a clearly marked and secured perimeter?
  - Is the car park well-lit in all areas?

Warnings and safety information
- Objective: Safety information will be prominently placed for guest information.
- Risk Evaluation:
  - Are lift lobbies marked with signs warning against use of elevators (lifts) during fire alarms?
  - Do you use "warning: wet floor" signs when cleaning floors with slippery surfaces?
  - Are all external doors on fire escapes marked with signs saying "Fire escape—do not block" or a similar text?

Security VIPs

- Objective: Guests will enjoy privacy during their stay.
- Risk Evaluation:
  - Do you have a specific policy for handling security VIPs?
  - Does planning for handling security VIPs include arrangements to ensure the safety and security of other guests and staff members?
  - Do you have a contact at the foreign office protocol department that can assist you with planning formalities?

Safe deposit

- Objective: No guest will suffer any loss during their stay.
- Risk Evaluation:
  - Does the hotel have safe deposit or in-room guest safes available 24-hours and free of charge to all registered guests?
  - Is there a procedure for always checking guest ID before allowing access to a safe deposit box?
  - Are in-room safes installed securely according to manufacturer's specifications?

Baggage storage

- Objective: Guests will not suffer any loss during their stay.
- Risk Evaluation:
  - Is the baggage storage room always locked and protected in a way that ensures access is controlled and monitored?
  - Do guests receive receipts for each piece of luggage stored?
  - Is storage available to the general public in addition to hotel guests?

Loss and theft

- Objective: Reported losses will be handled in a professional manner and reported to the relevant authorities for investigation.
- Risk Evaluation:
  - Does the hotel have a procedure for handling reported loss or theft?
  - Does the procedure for handling loss/theft include securing evidence (video recordings, lock readouts, etc.) that may be useful to investigating authorities?
  - Are all reported cases of loss or theft immediately reported to the manager on duty?

Video surveillance

- Objective: Hotels will limit the risk of unwanted activity.
- Risk Evaluation:
  - Does a data protection act or other form of legislation requiring you to obtain a license or register your system in any way govern the use of CCTV in your property?
  - Do signs inform anyone entering an area where camera surveillance is in use?
  - Is the surveillance data storage unit placed in a secure location?

Packages and post
- Objective: Packages and post will be carefully handled to ensure they reach the intended recipient without causing harm to operations or personnel.
- Risk Evaluation:
  - Do you have a procedure for handling mail, parcels, and couriered items?
  - Are mail and packages kept in a secure location until received by intended recipient?
  - Is mail only accepted for registered guests, guests with reservations, or employees?

Inspection of guest and public areas
- Objective: Guests will enjoy a safe and secure atmosphere during their stay.
- Risk Evaluation:
  - Does the hotel carry out regular inspections of guest room corridors and public areas to look for unauthorized persons or other potential hazards?
  - Are inspections documented?
  - Is there a system in place to ensure follow-up of inspection reports?

Restaurant, banquet, and meeting rooms
- Objective: Guests will enjoy a safe and secure atmosphere during their stay.
- Risk Evaluation:
  - Are your restaurants and meeting and event facilities approved for use by relevant local authorities?
  - Are facilities inspected by staff daily to ensure that fire exit access is clear and that all emergency equipment is in place and in good working order?
  - Do you have a policy for use of candles, open flames, and for disposing of ash or other potentially hazardous materials?

Solicitation and prostitution
- Objective: Guests will enjoy a safe and secure atmosphere during their stay.
- Risk Evaluation:
  - Do you have a policy forbidding staff to mediate for prostitution and solicitation?
  - Have you checked with the police in your city regarding guidelines on prostitution in hotels?
  - Are your bar and restaurant staff instructed to inform the manager on duty if there is a person suspected of prostitution in their outlet?

Suspected loss of life
- Objective: The integrity of a deceased person will be maintained.
- Risk Evaluation:
  - Does the hotel have a specific procedure for handling cases of suspected death?
  - Does the procedure include immediately informing head office?
  - Will all cases of suspected death in your hotel automatically lead to a police investigation?

Property damage
- Objective: Guests will enjoy a safe and secure atmosphere during their stay.
- Risk Evaluation:
  - Does the hotel have a special damage report procedure that ensures immediate contact is established to relevant authorities and insurance company?

- Does the hotel possess a camera that can be used to document evidence of damages?
- Are all staff aware that only the general manager may accept responsibility to cover the cost damages may inflict?

Handicapped guests
- Objective: All guests, including those with a handicap, will enjoy a comfortable, safe, and secure atmosphere during their stay.
- Risk Evaluation:
  - Have you asked local organizations to assist in planning how you best can cater to the needs of the handicapped?
  - Do you have policies and procedures for assisting guests with different forms of handicap (sight, hearing, movement, multihandicap, etc.)?
  - Do at least 1 percent of rooms have handicap facilities?

Accidents and illness
- Objective: The hotel will work to limit the risk of any guest becoming injured or ill.
- Risk Evaluation:
  - Do you have procedures for handling medical emergencies?
  - Are staff trained to follow specific procedures if a medical emergency arises?
  - Do you have procedures for following up medical emergencies to limit the chance of reoccurrence?

Employee safety and security
- Objective: The hotel will work to limit the risk of any employee becoming injured, ill, or suffering any loss at work.
- Risk Evaluation
  - Is safety and security included in the new employee orientation/induction program?
  - Has any safety and security training been carried out during the past 3 months?
  - Has a workplace environment inspection been carried out in all departments during the past 12 months?

Information and employee handbook
- Objective: Employees will be informed and trained on matters concerning safety and security in their workplace.
- Risk Evaluation:
  - Does each employee receive an employee handbook or other written document that includes rules, regulations, and safety procedures?
  - Does the hotel have a workplace safety committee that actively follows up safety and security incidents to ensure that accidents are prevented or prevented from reoccurring?
  - Do you carry out yearly inspections of each workplace to ensure that they are safe?

Locker rooms
- Objective: Staff will have access to hygienic changing rooms that include toilet and shower facilities and lockers for uniformed staff to keep their private clothes in.

- Risk Evaluation:
  - Do you have changing rooms for all staff?
  - Do changing rooms provide facilities such as showers, sinks, and toilets?
  - Do all staff with uniforms have access to a locker for their private clothes?

Work area inspections
- Objective: The hotel will work to limit the risk of any employee becoming injured, ill, or suffering any loss at work.
- Risk Evaluation:
  - Are employees trained to check daily the safety and security of their workplace?
  - Are documented inspections of workplaces carried out regularly?
  - Are employees aware of to whom they should report any faults or dangers?

Hazardous chemicals
- Objective: The hotel will work to limit the risk of any employee becoming injured or ill.
- Risk Evaluation:
  - Does the hotel have a register/safety data sheet (SDS) of any hazardous chemical used for cleaning, maintenance, or other purpose?
  - Do you require suppliers to provide datasheets containing information on content, how to handle spillage, misuse or injury?
  - Is protective equipment such as gloves or eye protection available?

First aid kits
- Objective: The hotel will work to limit the risk of any employee becoming injured or ill.
- Risk Evaluation:
  - Is there a system in place to ensure that the hotel always has first aid trained and certified staff on hand?
  - Does the hotel have an automatic electronic defibrillator (AED) and are staff trained to use it?
  - Do you have a system for checking and ensuring that first aid kits are kept properly stocked and refilled?

Accidents and illness
- Objective: The hotel will work to limit the risk of any employee becoming injured or ill.
- Risk Evaluation:
  - Do you have procedures for handling and documenting staff accidents or illness?
  - Are staff trained to follow specific procedures if a medical emergency arises?
  - Do you have procedures for following up medical emergencies to limit the chance of reoccurrence?

Key control and lock codes
- Objective: The hotel will work to limit the risk of any employee suffering any loss in the workplace.
- Risk Evaluation:
  - Do you have procedures for handling staff keys to ensure that only appropriate staff have access to areas necessary for their duties?
  - Do you keep a daily log of staff key usage?
  - Do you have a procedure to ensure that all master keys can be accounted for?

Entry and exit control
- Objective: The hotel will work to limit the risk of loss and theft.
- Risk Evaluation:
  - Do you have a dedicated staff entrance that must be used by all staff entering or leaving the workplace?
  - Do you carry out spot checks on staff to help protect against internal theft?
  - Do staff have uniforms, nametags, and ID cards for identification?

Crisis management organization
- Objective: With the goal of saving lives, protecting the brand and property, any disruption of operations will be handled professionally to limit the time and scale of the disruption.
- Risk Evaluation:
  - Do you have crisis management policy and procedures?
  - Can the crisis management team always be put into action at short notice?
  - Is the crisis management team trained and aware of corporate reporting procedures?

Fire safety
- Objective: All hotels will actively work to prevent fires and to reduce the risk of igniting fires on the hotel property. All hotels will have equipment, procedures, and training programs designed to limit the risk of damage done if a fire should occur. As a minimum, hotels must adhere to the fire regulations as defined in the state or country in which the hotel operates. In cases where the following practices may conflict with a local law or fire code, the local law or code must be followed.

External fire safety inspections
- Objective: As a minimum, hotels must adhere to the fire regulations as defined in the state or country in which the hotel operates.
- Risk Evaluation:
  - Do you keep a record of all fire safety inspections?
  - Do you keep a record of all follow-up carried out after fire safety inspections?
  - Are all necessary operating licenses and other mandatory internal documentation up-to-date and available for inspection?

Control of electrical installations
- Objective: All hotels will actively work to prevent fires and reduce the risk of igniting fires on hotel property.
- Risk Evaluation:
  - Do you keep a record of all electrical installation inspections?
  - Are staffed trained not to use faulty equipment and to report it for repair?
  - Do you have thermographic imaging done on major electrical connection points, fuse boxes, etc?

Equipment—fire alarm system
- Objective: All hotels will have equipment, procedures, and training programs designed to limit the risk of damage should a fire occur.

- Risk Evaluation:
  - Do you have a fully automatic, fully addressable fire alarm system?
  - Is the fire alarm system connected directly to the fire department or other external alarm-monitoring center?
  - Are staff trained to understand and immediately respond correctly to any indication of warning, alarm, or fault that is registered by the fire alarm panel?

Equipment—sprinkler
- Objective: All hotels will have equipment, procedures, and training programs designed to limit the risk of damage should a fire occur.
- Risk Evaluation:
  - Does a licensed company in accordance with manufacturer's specifications regularly test, maintain, and document the status of the system?
  - Have you checked to ensure that sprinkler effectiveness is not reduced by storage closer than 18″ (50 cm) to the sprinkler head?
  - Do you have procedures for limiting water damage in the event a sprinkler is set off?

Equipment—fire extinguishing
- Objective: All hotels will have equipment, procedures, and training programmes designed to limit the risk of damage should a fire occur.
- Risk Evaluation:
  - Is appropriate fire extinguishing equipment readily available in all areas of the hotel?
  - Are all extinguishers, hoses, and other such equipment tested regularly and is there documentation of these tests?
  - Are all staff trained in the proper use of fire extinguishing equipment in their workplace?

Compartmentation
- Objective: All hotels will have equipment, procedures, and training programs designed to limit the risk of damage should a fire occur.
- Risk Evaluation:
  - Do you have up-to-date documentation showing where each fire cell is?
  - Are staff trained to check fire doors in their workplace daily?
  - Do you conduct regular tests of fire doors and document that they are unobstructed and in good working order?

Fire escapes
- Objective: All hotels will have equipment, procedures, and training programs designed to save lives should a fire occur.
- Risk Evaluation:
  - Does the hotel have multiple fire escapes from each floor providing fire protection until one is completely outside the building?
  - Are all fire exits marked with illuminated signs that are connected to a back-up power source to ensure they remain lit during a power failure?
  - Can you document that fire exits are checked daily to ensure they are unobstructed?

Emergency lighting
- Objective: All hotels will have equipment, procedures, and training programs designed to save lives should a fire occur.

- Risk Evaluation:
  - Do you have emergency lighting that covers public areas, back-of-house areas, corridors, and emergency stairwells?
  - Is emergency lighting tested at least four times per year?
  - Do you have a documented maintenance program that shows testing, battery changes, etc.?

Internal fire prevention inspections
- Objective: As a minimum, hotels must adhere to the fire regulations as defined in the country in which the hotel operates.
- Risk Evaluation:
  - Do you carry out regular internal fire safety inspections?
  - Do you keep a record of all internal fire safety inspections?
  - Do you keep a record of all follow-up carried out after internal fire safety inspections?

Avoiding common ignition sources
- Objective: All hotels will actively work to prevent fires and reduce the risk of igniting fires on hotel property.
- Risk Evaluation:
  - Have you issued a smoking policy that only allows smoking in designated areas?
  - Do you store all flammable liquids and spray cans in a safe way (e.g., inside a closed metal cabinet) when not used?
  - Do you issue written work permits for external contractors work, especially temporary hot work, such as welding, cutting, grinding, and mending the roof?

"Manager on duty" program
- Objective: Every hotel should always work to actively prevent operational disruptions in the hotel and to ensure that any disruptions that may occur are as limited as possible in cost, length, and degree.
- Risk Evaluation:
  - Do you have a policy that specifies one manager or employee is in charge of operations at any given time ("manager on duty" program)?
  - Are employees with "manager on duty" status fully trained in all aspects of hotel operations, policies, and practices?
  - Do all employees with "manager on duty" status have access to documentation regarding necessary operational and all emergency procedures?

Contingency plans
- Objective: Every hotel should always work to actively prevent operational disruptions in the hotel and to ensure that any disruptions that may occur are as limited as possible in cost, length, and degree.
- Risk Evaluation:
  - Do you have back-up and fallback routines that can be implemented when any operational system fails or when other factors threaten normal operations?
  - Are contingency plans included in the "manager on duty" documentation so they are readily available for implementation at all times?
  - Can you document that staff with "manager on duty" status are trained in implementation of contingency plans for system failures and operational disruption?

Evacuation
- Objective: Every hotel should always work to actively prevent operational disruptions in the hotel and to ensure that any disruptions that may occur are as limited as possible in cost, length, and degree.
- Risk Evaluation:
    - Is the evacuation alarm tested regularly?
    - Do you keep records of tests and maintenance of the evacuation alarm?
    - Have you carried out a full-scale evacuation drill within the past 12 months?

Emergency evacuation kit
- Objective: Every hotel should always work to actively prevent operational disruptions in the hotel and to ensure that any disruptions that may occur are as limited as possible in cost, length, and degree.
- Risk Evaluation:
    - Do you have an evacuation kit?
    - Do you have a system for registering all guests, employees, and others who are evacuated?
    - Do you have an agreement with another building to ensure that evacuees can be relocated to a safe building as soon as possible after an evacuation?

Internal control
- Objective: All hotels shall maintain a structured system to ensure that safety and security measures are followed on a daily basis in normal operations.
- Risk Evaluation:
    - Have you completed or updated your Carlson Fire Safety Survey form within the past 12 months?
    - Have you completed and updated the Carlson Rezidor Safety Security Self-Assessment for each quarter of the year?
    - Does your hotel have policies, procedures, and training to document safe working practices are in place?

Internal control—food safety
- Objective: Every hotel should always work to actively prevent operational disruptions in the hotel and to ensure that any disruptions that may occur are as limited as possible in cost, length, and degree.
- Risk Evaluation:
    - Do you have a food safety program (e.g., hazard analysis and critical control points [HACCP])?
    - Does your food safety program include all points listed in the best practice example?
    - Does your food safety program include documenting all parts of the food safety chain from supplier to end user?

Internal control—water safety
- Objective: Every hotel should always work to actively prevent operational disruptions in the hotel and to ensure that any disruptions that may occur are as limited as possible in cost, length, and degree.

- Risk Evaluation:
  - Do you have a water safety program?
  - Do you conduct regular tests for legionella?
  - Does your water safety program require treatment such as heating hot water boilers to 60°C (140°F) to prevent waterborne disease?

Internal control—ethics

- Objective: The Code of Ethics and Business Conduct will serve as guidelines for our business conduct and responsibilities vis-à-vis colleagues, customers, guests, suppliers, shareholders, authorities, and the world at large.
- Risk Evaluation:
  - Are all employees informed and educated in what the Code of Ethics and Business Conduct covers?
  - Are all employees informed of how they should report concerns and breaches of the items covered in the Code of Ethics and Business Conduct? Do you have procedures in place for how to handle reports or allegations concerning possible breach of the Code of Ethics and Business Conduct?

Crisis management program

- Objective: Every hotel should always work to actively prevent operational disruptions in the hotel and to ensure that any disruptions that may occur are as limited as possible in cost, length, and degree.
- Risk Evaluation:
  - Do you have a crisis management program?
  - Does your crisis management program include procedures for how and when to notify Carlson Rezidor?
  - Have you conducted crisis management training within the past 12 months?

Traveler's perspective

As risks posed to travelers evolve, and travelers continue to travel to new and increasingly dangerous destinations, safety training with recurring updates are a worthwhile investment for employers financially, and time-wise for travelers. There are some common themes to hotel safety tips for all travelers. However, while some tips and safety precautions are universal, there are some that are specific or more applicable to certain locations around the world or to certain types of travelers. The following safety tips are broken down into categories from the traveler's perspective by location or traveler type.

General traveler hotel safety tips

1. Try to avoid using hotel Wi-Fi unless you are using a company approved VPN (virtual private network) that can encrypt information going out of and coming into your computer. Recent U.S. Federal Communications Commission (FCC) rulings have made it illegal for

hotels to block Wi-Fi signals other than their own, which some major chains were caught doing and were fined accordingly. If no VPN is used, try and use your mobile device as a hotspot for Internet access if available.

2. Never conduct financial transactions while using a hotel or public Wi-Fi service.
3. If traveling alone, only ask for one hotel room key. A second key lying around in your room invites housekeeping to take it and allow himself or herself or someone else to easily get back in later and rob you.
4. Check all of the locks on the doors and windows of your room, to make sure that they work, and contact the front desk if they do not.
5. Make sure that the hotel phone is working, and that you know how to make both internal hotel calls and external calls.
6. Make sure that you know the local emergency services phone numbers for fire, police, and ambulance services.
7. Never allow your front desk agent or other hotel employees (e.g., restaurant employees) to say your room number aloud in the presence of other guests. When communicating a room number, write it down.
8. Know your nearest fire escape route.
9. Avoid hotels with outside access to the room door (e.g., many motor lodges or motels have rooms where you can see the door from outside). Female travelers traveling alone should take great care to only stay in hotels where the entry to their room is inside of a secure facility with on-duty personnel (either front desk and/or security).
10. Always stay with your luggage, when checking in or checking out. Use a bellman, if necessary.
11. If you are arriving by personal or rental car, try to use valet parking. If unavailable, park as closely as possible to your hotel point of entry.
12. Avoid dimly lit parking garages. They are often not monitored and are without supervised security.
13. When possible, select hotels with security "peepholes" in the doors to see who is on the other side. However, make sure that the view access from the inside has a cover for access, because there are devices that allow people from the outside to get a broad view of the inside of your room from the outside unless covered/blocked. Otherwise, make sure that you have a device to block the view from the inside. In 2011, an ESPN reporter filed a $6 million dollar lawsuit against a hotel where a man filmed her from inside her hotel room via the peephole in the door. Peepholes, can be useful for guests, but travelers should be aware of the potential for risks as well.
14. Ask what kind of physical security the hotel provides and where?
15. Ask if all employees have had background checks and are drug tested.
16. Are all rooms equipped with fire safety smoke alarms that are connected to the central fire alarm?

Hotel security tips for high-risk destinations

1. Use of hotel safes aren't 100 percent secure, so bear that in mind while traveling and consider travel security products that can be affixed to permanent fixtures or furniture in the room. There are a variety of new and interesting travel security products on the market to consider with features like slash proof fabric and cables or straps that are difficult to cut. Avoid bringing anything of significant value with you, such as jewelry or large amounts of cash.

2. Supplement your hotel door locks if you can, with a traveler door jam or similar device. These devices can provide you with precious additional time when an assailant tries to break into your room with you in it!

3. Stay at hotels with restricted access for vehicles to the front hotel lobby doors. Any access to the lobby doors from public or private vehicles must go through a security check that requires vehicle searches prior to advancement to the driveway.

4. Proximity between outside vehicles and the lobby doors should be significant in consideration of a potential blast radius.

5. Understand what level of security is provided at your hotel and how to contact them in an emergency.

6. Preprogram emergency contact phone numbers in your mobile device prior to arrival (crisis response hotline, local emergency services, consulates and embassies, etc.), or utilize your travel risk management (TRM) provider applications that connect you with crisis response support services with a touch of a button.

Hotel room access, benefits, and features for travelers via mobile devices

In 2015, major hotel chains will allow travelers to use their phones to select their rooms, check in and out, receive special push notifications such as potential happy hour specials, and to use their mobile devices to gain entry to their rooms, bypassing the front desk.

Hilton Worldwide has reportedly invested $550 million dollars in 2014 for some of these mobile features to be used across multiple hotel brands within their portfolio. The jury is still out as to the effectiveness and safety of using mobile devices for keyless hotel room entry, but to some extent, there have always been security risks associated with how travelers access their hotel room. Even with old-school metal keys, there was always the possibility for duplicates to be floating around in the general public or with hotel staff. However, as password technology evolves, and randomly assigned, complex passwords are easily managed for each program that business travelers sign up for, such as with Apple's "keychain access," such technologies show considerable promise.

Conferences, meetings, and incentive trips

8

Group travel for conferences, meetings, and incentive trips aren't managed from a risk perspective in exactly the same way as individual business trips. A primary reason for this is that the group travel reservations data is structured differently than is the data for individual, or what we also call "transient," travel and is often managed in other systems that are designed for working with and managing multiple groups. Even though some aspects of group travel, such as air transportation, does end up in a GDS (global distribution system) such as Sabre or Amadeus, group reservations are not easily parsed or accepted by traveler tracking platforms. In short, group travel and transient travel aren't generally tracked in the same systems because of these data challenges, which forces meeting and incentive managers to track travelers and manage risks separately, often manually. The challenge comes with how group passenger name records (pnrs) or reservations are formatted versus single traveler pnrs or reservations. With a single traveler passenger name record, you have one name, and each segment applies to that one name in the reservation. With a group passenger name record, you could have 15 names (for example) and still one or more segments for air transportation, and it applies to all 15 names in the one reservation. Breaking these out into individual traveler database entries for travel risk management (TRM) solutions has historically been a challenge, but is something that the industry is working to address.

Event management platforms, are ideal sources for group travel data if you manage your meetings and events through them, because they consolidate traveler and event data into one place, which is predicted to someday be the foundation for integrating group and transient travel data into traditional traveler-tracking TRM systems.

Data and tracking aside, meeting and event security at a macro level should consider the following precautions:

1. Preevent venue safety and security inspection
2. Impact of the agenda on security coverage
   a. Movement of attendees and security personnel
   b. Risks associated with signage at the event
3. Policy compliance for attendees and security personnel (where applicable)
   a. Venue new hires (collaboration between hotel security and private security, or other personnel)
   b. Background checks for all vendor and security personnel
   c. Documented and disclosed security concerns, guidelines, and policies about potential risks relative to the venue location and types of activities associated with the event (e.g., some types of team-building exercises)
4. Positioning of security personnel throughout the event
5. Communication plan for attendees and security personnel, including instructions in the event of an emergency, resource availability, and support

6. Crowd control and coordination during an emergency
   a. In the event of an emergency, what technology or mechanisms do you use in order to communicate with all participants?
   b. Emergency ground transportation or ability to shelter in place
7. Is the event venue indoors or outside?
   a. How does this impact your potential risks?
   b. How does this impact your potential security mitigation plans?
8. Access to adequate medical support for individual or multiple injuries on a large scale. Of course, there are many more potential precautions to take, but those provided represent only a sampling of them.

Group air transportation

Important factors to consider for managing air transportation for groups include:

1. How many employees total do you have on any one flight?
   a. How many executives do you have on any one flight?
   b. What percentage of any one department do you have on any one flight?
2. What is the safety rating and/or maintenance history of the airline you are considering using?
   a. Have you checked the airline and its associated CAA (civil aviation authority) for their safety ratings and records with organizations like IOSA (IATA Operational Safety Audit).
3. Is it safer to charter private transportation? While in general, it is best to keep a minimal number of employees on the same flight, in some parts of the world, where there are few options for commercial air transportation, some companies pay more money for private transport from suppliers that maintain a higher standard of safety and maintenance than is available from commercial air transportation. These instances should be managed in consideration of, and in concert with, a company's security and legal departments, and are typically considered for recurring group movements versus one-time charters, and are highly supervised.

Austin, Texas–based Freescale Semiconductor allegedly had 20 people on Malaysian Airlines flight 370 that went missing on March 8, 2014, somewhere over the South China Sea. The important thing to note is that we don't know if this was organized "group travel" or simply 20 individually booked business trips from the same company that happened to be on the same flight. Either way, the identification and monitoring of a large number of employees on the same flight must be done prior to travel in order to properly mitigate the potential risks, such as by forcing some travelers to rebook onto other flights. Some travel agency reporting tools, or TRM platforms, can set up rules to trigger a report when a maximum number of employees per flight is exceeded. Many companies have two policies on this topic, which includes a maximum number of employees per flight overall, and a second policy for a maximum number of executives on the same flight (i.e., VP and above).

Aside from the emotional toll and loss-of-life impact on the company, family, and friends, imagine the impact to the organization that loses 20 employees on the same flight. What if an entire project team or department was lost? Such a loss could kill a project, cause loss of revenue that results in job losses, and even cause damage to the company's reputation. In today's job market, many potential new hires are asking questions about how prepared employers are to protect them from unnecessary risk and how well the employers manage incidents when they happen. This is a good example of how TRM, can have a direct impact on a company's operational risk plan, which is directly tied to productivity. That is how and why companies need to make considerable investments and plans with regard to overall risk management, including travel, operations, assets, supply chain, and expats.

Hotel location and venue selection

From a risk perspective, the following aspects of hotel and venue selection for conference, meeting, and incentive travel should be considered:

1. Where are they located?
   a. How close are they to emergency services?
      i. Police
      ii. Fire departments
      iii. Ambulance and hospital support
2. What kind of venue are you using in conjunction with the meeting or event?
   a. Hotel
   b. Conference center
   c. Concert hall
   d. Shared economy suppliers (e.g., Airbnb or Homeaway) for sleeping accommodations
3. Can the local infrastructure and emergency services support your group size in a timely manner?
   a. For large groups, what assets are readily available on a moment's notice, particularly for incidents involving large numbers of people?
      i. Do you need to hire your own emergency services infrastructure for the event (security and medical services)?
4. Health, safety, and sanitation information
   a. Does the hotel or its facilities have any health and safety ratings?
   b. What kind of food safety and sanitation training is required of hotel employees?
   c. Is the tap water drinkable, or should you recommend the use of bottled water only?
5. Is the building or venue up to fire and safety code standards?
   a. Recent licensing or evaluations?
6. Is the building or venue handicap accessible?
   a. Sleeping rooms
   b. Meeting rooms
   c. Restaurants
   d. Common areas
7. Security
   a. Have all employees undergone thorough criminal and drug background checks?
      i. How often are they rescreened?
   b. Does the building or venue provide security services? If so,
      i. Where?
      ii. How many personnel?
      iii. Hours of available service?
      iv. How to request support?
      v. Can they support customer requirements?

    c. Do any sleeping rooms provide deadbolt and secondary locks?

    d. Does the building or venue have controlled access after hours?

    e. Are elevators equipped with key-controlled security access?

    f. Have you researched local crime statistics within a 2-mile radius of the building or venue?

8. Natural disasters

    a. Evacuation plans or shelter in place for

        i. Hurricanes

        ii. Tornados

        iii. Floods or tsunamis

9. Fire safety

    a. Are emergency fire evacuation plans posted?

        i. Where?

    b. Are there sufficient fire extinguishers available throughout the building or venue?

    c. Are there fire alarms in every room, common area, and meeting room?

        i. Are they connected to a central alarm system that contacts the local fire department?

10. Terrorism

    a. Are there plans in place to deal with potential events such as:

        i. Active shooters

        ii. Kidnapping/hostage situations

        iii. Bombing

11. Insurance

    a. Does the building or venue have current and sufficient liability insurance?

12. Building or venue maintenance

    a. How often is the facility renovated?

        i. When was the last renovation?

            1. What was included?

    b. Are all elevators and escalators current with safety inspections?

        i. Do they have alarms with emergency calling access?

        ii. Do they contact the authorities when alarm is triggered?

        iii. Are the elevators equipped with security cameras inside that record activity?

Insurance

1. Employer/host insurance

    a. Have you verified that your Business Travel Accident (BTA) policy covers the event, especially if located in an international location?

    b. Do you have a standard amount of coverage for meetings and events?

    c. What gaps are there in insurance coverage needed for the event?

        i. Will employee or attendee's private health coverage provide coverage for the location, or only subjective reimbursement after services and/or medical treatment is provided?

        ii. Are any special activities, such as team-building events that are considered "extreme sports" or dangerous in any way?

            1. Pool parties with alcohol

            2. Zip lines

            3. Ice skating

            4. Anything with the potential for injury

        iii. What types of ground transportation is covered for attendees?

    d. Have you notified your insurance providers of your meetings and events, along with any special activities?

Insurance considerations for meetings and events

When planning a meeting or event, there are several aspects of managing insurance that need to be considered.

1. While sometimes insurance can be costly, have you done a cost-to-benefit analysis to see which risk is the least difficult to manage? (a) The cost of the premium, which you should include the overall cost of the event, or (b) the cost of canceling the event and losing deposits in addition to any potential claims or lawsuits that may arise unexpectedly?
2. What kind of coverage do you need?
   a. Convention cancelation coverage
   b. Commercial general liability coverage
   c. Basic properly insurance
   d. International property and casualty insurance
   e. Business auto policy
      i. Include uninsured motorists coverage
      ii. Property damage for the vehicle driven by an employee or contractor
      iii. Liability coverage for additional vehicles involved in accidents with an employee or contractor
      iv. Medical coverage for those involved in the incident where liable.

Always consider the many potential exclusions to insurance policies considered for coverage of meetings and events. It is especially easy to assume that BTA policies or corporate credit cards used for the rental of cars or ground transportation provide adequate coverage, yet sometimes they do not. Often specific locations, types of vehicles, or other circumstances will limit or exclude coverage.

Insurance for meetings and events isn't and shouldn't be just to recover losses for cancelations, but also largely to provide funds should the need arise to defend against lawsuits. By having in place standard safety and security processes and protocols for different types of emergencies, companies can reduce their liabilities by showing a compelling effort to keep participants safe, and potentially save money in insurance premiums, depending upon how your provider rates you based upon your policies and programs. Such risk management strategies and programs have been known as useful negotiation tools with providers.

Ground transportation

1. What kind of vehicles or other means of ground transport are being used or provided to meeting attendees?
   a. Public transportation
   b. Private chauffeured car services
   c. Car rentals
   d. Buses
   e. Trains
   f. Taxis
   g. Shared economy transportation

Meetings planners

Whether making travel arrangements for an individual, or organizing logistics for multiple people to attend a meeting or event, companies and, indeed, the general public think that simply because they are capable of making a simple reservation of some kind, that they no longer need the support or experience of an experienced travel agent or meeting planner. What they don't understand, aside from this being a horrible untruth, is that with making reservations come great responsibility and the potential for liability. This is why companies shouldn't let just anyone plan and book meetings, even small ones, without appropriate training and the assistance of trained professionals who can ensure that a minimum degree of precaution and care is taken to address safety and security standards relative to both organizing a meeting or event and to precautions during the meeting or event itself.

From small meetings to large events, not all attendees are employees of the meeting or event sponsor, and may not be easily tracked or incorporated into a TRM solution or platform that enables traveler tracking and communications. Planning and consideration should be given to how all meeting and event attendee itinerary and contact details (in particular, mobile phone numbers and e-mail addresses) are collected in such a way that they can be used both before and during the event to provide appropriate risk disclosures or crisis communications. Sometimes these communication tools for meetings and events are managed separately from TRM solutions, in order to facilitate sending event-related, nonsecurity messages to attendees or participants, but what is most important is that everyone can be seamlessly communicated with by meeting and event organizations and security teams. Ideally, TRM systems would have all of the relevant information for all meeting or event attendees that booked travel via the employer's managed travel program, or potentially having collected the details from those who didn't via an open booking application. In parallel, the meeting and event platform should have that same participant contact and itinerary information. Ensuring that all attendee's contact and itinerary details reside in both a meeting/registration solutions platform and a TRM solutions platform, respectively, provides an ideal foundation for managing both the details, agendas, changes, and standard communications with meeting and event participants, while the TRM tool can automatically send out any relevant security briefings or alerts, or crisis communications on an ad hoc basis. Employers may or may not choose to use multiple platforms in parallel when it comes to meetings and events, but if they do, it is important that travelers understand and expect that they may receive communications from disparate systems. While it may not sound practical to use multiple systems, bear in mind that the way that transient travelers are managed from a risk perspective, is different from how expatriate travelers are managed. Accordingly, meeting and event attendees require some of the same disclosures and features from a TRM solution as a transient traveler, but they also require additional support specific to the meeting or event that they are attending.

Even with the use of a professional third-party meeting planner and/or TRM firm, it is the employer who owes a duty of care to those traveling to and/or attending meetings, events, and conferences to:

1. Investigate the hotel or venue space to ensure that it meets employer or industry safety standards for building safety, security, sanitation, and emergency response readiness, including background checks for employees.
2. Disclose any known or potential hazards or safety precautions to attendees or participants and reiterate company policies and codes of conduct with regard to things such as excessive consumption of alcohol or restrictions on events or team-building activities that may be considered extreme or dangerous. Potential document distribution should include information on secure ground transportation, emergency services, and contact information for medical attention or the police, emergency and fire evacuation plans, and a contracted third-party crisis response hotline (when applicable).
3. Prior to meetings and events beginning, all stakeholders (attendees, venue and security staff, TMC [travel management company] support, etc.) should be made aware of an established communication plan for both event-related information (printed and/or electronic) and crisis communications. Real-time communication with attendees and staff via meeting/group SMS messaging applications can be extremely helpful with keeping people informed of changes to meeting agendas, as well as developing or emerging risks.
4. Execute measures to avoid exposure of risks to attendees or participants, such as physical security resource utilization (if applicable), entry and exit restrictions for the meeting or event, and containment of attendees throughout the event as best as reasonably possible within a designated and secure environment. An example of when this might be needed is if an event is held in a city that is rated with a high or very high risk rating.

Ask yourself if in planning your meeting, event, or conference, whether you have considered emergency planning and response and/or insurance coverage for any or all of the following:

1. Medical assistance and evacuation for all participants in the meeting or event, including vendors or suppliers;
2. Trespassers or "crashers";
3. Natural disasters;
4. Riots or civil unrest;
5. Acts of terrorism (biological or chemical attacks, or active shooters).

Having sufficient security personnel onsite with constant contact and compliance to program guidelines is essential. Security personnel should be briefed in advance and onsite at the event, because one breach in security can disrupt the event, causing evacuations and even physical harm to one or more guests. For larger events with physical security, evacuation plans should consider how any confidential, intellectual property that may have been left behind in the event of an evacuation is to be safeguarded. Who secures the property or disposes of it?

White Paper: Keeping High-Profile Meetings Safe and Secure

by Harlan Calhoun

Before the advent of smart phones, planting covert listening devices was the most popular way to illegally record content from a private meeting. Today, with an estimated 130 million smartphones in use in the United States, every user has the potential to be a covert meeting operative with their own Wi-Fi receiver, camera, audio recorder, and keyboard and computer at their disposal.

Maintaining security and privacy for high profile meetings is vital whether it is a shareholder meeting, political fundraiser, or internal executive meeting. When off-the-cuff comments made by Mitt Romney at a fundraiser were recorded and posted online, his political campaign took a negative hit. The theft of intellectual property costs American businesses billions each year. Discussions of proprietary information may include customer pricing, R&D (research and development) and production processes, marketing and advertising strategies, legal issues, and salary information. How can corporate and government leaders ensure that their meetings are safe and secure and that proprietary information is not leaked?

As a security and law enforcement professional, I have been responsible for the safety and security of Fortune 500 executives and have seen the positive results of a well-planned meeting. These guidelines can help establish processes and protocols to keep people and information protected.

Preplanning is critical

About 3 months before the meeting, all critical parties (human resources, legal, internal security, external security, local law enforcement, and facilities) should meet to begin the planning process and assess the level of risk associated with this meeting. Questions to ask include: how many people are expected, where the meeting is being held (offsite or at corporate headquarters), and what are the parking logistics? Will there be sufficient security personnel assigned to the meeting? What is the training of these officers? Who would have an interest in the content of the meeting? How could this content be utilized?

While corporate locations generally already have significant security controls in place, an important meeting could attract crowds or protesters or the content of the meeting may be such that additional security measures are required. Additionally, offsite meeting locations, such as hotels and convention centers, demand more advance planning, so that each area to be accessed is reviewed and accounted for from a security perspective.

Effective preplanning also includes social media and news monitoring to assess what topics are prominent in the news and how the individual company may factor in. For example: if it is an energy company, how are their environmental practices interpreted by organizations such as Greenpeace, and what impact might that have? Active social media monitoring of Twitter, Facebook, Instagram, and other channels can help intercept and prevent potential flash mob scenarios.

One month before the meeting, a tabletop exercise should be conducted where everyone is tasked with creating potential meeting crisis situations such as a medical emergency, disruptive attendee, or water main break.

Establish security protocol

Access control is critical and every meeting attendee should be thoroughly vetted, ensuring their identity matches their state or federal identification. The security firm should engage in an electronic sweep of the meeting room right before it starts to ensure there are no surreptitious listening devices planted. Metal detectors at the point of meeting entry should be employed.

All meeting attendees should be aware that security protocol is in place, which could include:

• *Pocketbook and briefcase check*—Smartphones, computers, iPads, and video and audio recorders should be removed and stored in a safe location.
• *Corporate computer used for presentations*—Speakers should provide their PowerPoint presentations in advance and all presentations should be run off of one master encrypted computer that is prescreened for bugs.
• *Postmeeting material check*—There should be a review of all written material taken by meeting attendees to ensure that no sensitive information leaves the meeting room.

Attendees should be briefed by the security team on appropriate protocol both during and after the meeting. Many secrets have been divulged at the hotel bar or gym when executives are not aware that competitors are located at an adjacent bar stool or treadmill.

Security protocol should also account for meeting disruption. A shareholder who owns just one share of a company's stock can lawfully gain access to a shareholder's meeting. It is not uncommon for an activist to purchase a solitary share for this reason. While every shareholder has a right to ask questions during a shareholder's meeting, no one has the right to be purposefully disruptive. It is important to preplan with corporate security and the company's public relations department to establish protocols to handle a disruptive questioner, or even uninvited media.

Meeting monitoring

Security and surveillance should remain in full force throughout the meeting. Security officers should conduct continual perimeter checks of the surrounding areas. There should be electronic and physical monitoring during the meeting. Make sure that all unencrypted wireless microphones, which can transmit meeting content outside of the room, have been removed and replaced with encrypted ones.

Escape clause

Sometimes even the best-laid plans can result in an unexpected surprise. A meeting that becomes unruly, for example, may require that the CEO and other top executives are able to depart the premises safely and quickly. Advance of the meeting, all escape routes should be detailed along with vehicles staged with drivers who can whisk away executives at a moment's notice. If it is not possible to get the executives out of the building, it should be determined what the shelter-in-place plan is and that this temporary shelter includes sufficient food and water.

Effective meeting security must also consider the health and welfare of all meeting attendees. If an attendee has a medical emergency, can the company access medical personnel quietly and without public incident? A CEO of a publicly traded company who suffers a heart attack, for example, should not be wheeled out to awaiting ambulances at the main entrance where people can observe his condition. Crowds can impede departure and news of a CEO's negative health issues could adversely affect stock price.

Confidentiality for all
Everyone who comes into contact with the meeting personnel (event planning personnel, security, audiovisual technicians, foodservice, administrative personnel, etc.) should sign a confidentiality agreement on behalf of the corporation and/or hosting hotel or venue. This should include not just those assisting the day of the event, but those involved in preevent preparations as well.

Postmeeting review
Conducting a postmeeting analysis is a valuable way to identify and document lessons learned from the project. The security and administrative team benefits from this postmeeting review of protocol and processes and is better equipped and prepared to face the next big meeting.

Meetings should not be publicized beyond the scope of meeting attendees. CEOs and other executives traveling to the meeting should be met by a driver who does not display the company name on their sign. There should not be any signs in the meeting location that draw attention to the nature of the meeting. Attendees should all understand that information should not be disseminated on any public Wi-Fi and that only secure, encrypted networks should be used.

These important events should never be an opportunity for corporate spies to gather proprietary information and intelligence. With proper planning, training, and resources, all meetings can be safe, secure and productive.

About the author
Harlan Calhoun is Vice President, Operations for AlliedBarton Security Services, www.alliedbarton.com, the industry's premier provider of highly trained security personnel to many industries including commercial real estate, higher education, healthcare, residential communities, chemical/petrochemical, government, manufacturing and distribution, financial institutions, and shopping centers. Prior to joining AlliedBarton, Harlan was a police officer, hostage negotiator and academy instructor in the Charlotte-Mecklenburg Police Department.

Source: Harlan Calhoun, "Continuity e-Guide: In the Know... Keeping High Profile Meetings Safe and Secure," *Disaster Resource Guide*, http://www. disaster-resource.com/index.php?option=com_content&view=article&id=224 7:keeping-high-profile-meetings-safe-and-secure&catid=8:facility-issues.

Enterprise risk management and its relation to travel risk management

9

Enterprise risk management (ERM) conceptually encompasses all manageable or potentially mitigatable risks that can impact personnel or business continuity, from travel and expats, to assets like facilities, supply chain and more. Some of methods for mitigation may be based upon regulatory requirements, ethical or environmental guidelines, credit or investment considerations, cyber risks, reputational risk, or compliance with safety standards. When approaching the topic of ERM alone, it can often be referred to in the context of finance-related decisions, but whether it is travel risk management (TRM), or any other aspect of risk management, safeguarding against these potential risks has value to the company and requires advance planning and systems to effectively reduce the potential for incidents or loss. From business travelers and the contributions they make, to a risk assessment in consideration of investing in an acquisition, at the end of the day, there is a financial value to approaching risk. However, in the case of TRM, there is also a human value, a personal impact on company travelers, and not just the value of their lives by keeping them safe, but the value of their levels of morale and stress, as it relates to how well they perform for their employers.

Putting enterprise risk into more context with TRM, consider all of the different ways that employee and contractor mobility touches most aspects of a company's operations, including a company's approach to risk management. For example:

- Travel to assess the risks associated with expanding operations into a new market.
- Travel to meet with parties being considered for merger, acquisition, or partnership, and their impact on the company's reputation.
- Travel to ensure compliance with legal and environmental requirements involving a project or operation
- Are your facilities still operational after a major natural disaster? What impact does that have on travel?

While the primary goal of this text is to develop an understanding and approach toward TRM as a discipline, and ongoing practice for process improvement, TRM is only one important element of an organization's ERM approach. Again, the areas managed under the guise of a risk management program can depend upon the lengths to which an organization has adopted risk management as a discipline.

Key components of ERM include:

1. TRM
    a. Employees
    b. Contractors
    c. Meeting and events
    d. Expatriates

2. ARM (asset risk management)
   a. Facilities
   b. Supply chain
   c. Intellectual property
3. ORM (operational risk management)
   a. Physical security of operations
   b. Resiliency of operations faced with interruptions (business continuity)

Components of TRM programs have distinct differences worth noting, such as:

- *Employee travel*—The primary driver and focus of most policies, plans, and protocols, for which the TRM program is built upon. This component most often takes the most time and resources pertaining to TRM.
- *Contractor travel*—While many of the same processes and programs that touch each of the key process areas of the Travel Risk Management Maturity Model (TRM3) apply to contractors, not all policies and procedures that apply to normal employee travelers necessarily apply to contractors. Whether a policy or procedure applies to a contractor depends upon many factors, such as the contractor contract requirements, contractor travel being booked outside of your travel program but paid for by your company, and the inability to provide same level of safety training or policy compliance with contractors. Thus it is important that there be clear terms and conditions for contractor travel and definitions of liability where possible (employer versus contractor).
- *Meeting and event attendee travel*—Management of this type of travel under TRM is unique because of the type of group transportation involved, as well as the types of travelers (employee and nonemployee attendees) and special safety considerations for managing risks at meetings and events, such as assessments, special insurance coverage, and support services (e.g., security).
- *Expatriates*—These employees are not managed the same as standard transient travelers, except when traveling away from their assignment on business. Typically, these individuals and their families are living in places for extended periods for business purposes, and may need a different kind of training, disclosures, and support for their assignment areas than transient travelers. For example, an expat living in Russia and traveling within Russia on business, may not want to receive the same security trip briefings over and over again, for trips that they make on a regular basis and have been trained already with regards to the associated risks and precautions. You may, however, want to continue forwarding relevant alerts to them for travel booked to any destination that they plan on visiting, but current updates on potentially critical incidents that may be cause for concerns over safety.

Unfortunately, expats can often get caught up in company-related business in foreign assignment countries, putting them in harm's way very uniquely because of their status or mere presence in the country in questions. For instance, in 2012, a Brazilian court required the surrender of passports from 17 employees from Chevron and oil rig operator Transocean, while the courts prepared criminal charges against the companies for alleged oil spill incidents.[1] Whether or not the employees were directly responsible for the incident, the Brazilian government allegedly thought that allowing these individuals to leave the country might negatively impact the Brazilian investigation.

---

[1] Simon Romero, "Brazil Bars Oil Workers From Leaving After Spill," *The New York Times*, March 18, 2012, http://www.nytimes.com/2012/03/19/business/energy-environment/brazil-bars-17-at-chevron-and-transocean-from-leaving-after-spill.html.

Another expat issue can include the fact that some countries refuse to allow residency to expatriates who have chronic health conditions such as human immunodeficiency virus (HIV), and require medical exams for applicants.

ARM is the structured ability to monitor things of value to an organization, from theft of interruption of delivery on supply chain, to a natural disaster's impact on office or manufacturing facility assets, among other incidents.

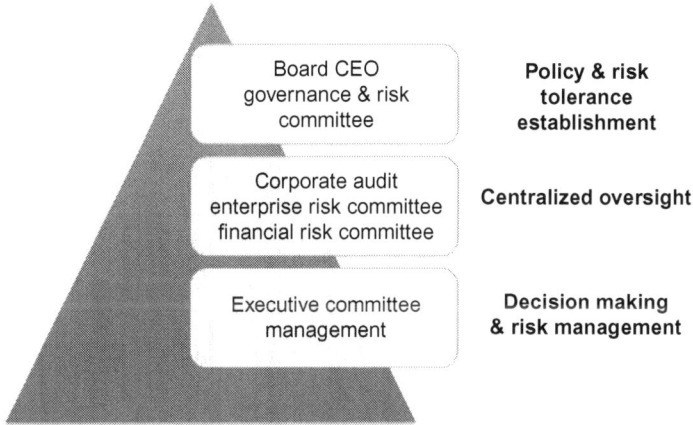

Board CEO governance & risk committee	**Policy & risk tolerance establishment**
Corporate audit enterprise risk committee financial risk committee	**Centralized oversight**
Executive committee management	**Decision making & risk management**

Example components of ERM governance.

Let's take a look at some different examples of ARM.

XYZ auto parts

XYZ Auto Parts is a fictional automobile parts manufacturer with factories and distribution centers throughout the southeastern United States, China, and Mexico. Mr. Vernon Smith is the company's risk officer in charge of ERM, including ARM.

Scenario 1–Supply chain loss investigation

Over the course of a 6-month period, an increasing percentage of losses have been experienced upon taking inventory of parts shipments to their manufacturing facilities in China. What didn't exist in the prior 6 months, has developed as a pattern over the most recent 6 months, and no one seems to know how valuable assets are turning up missing between their suppliers and their manufacturing facilities. Because the shipments travel through various shipping providers and in some cases internal distribution centers, the need to identify where the thefts are taking place is critical to solving the problem and ensuring that production isn't impacted by the shortage of supplies.

Similar to traveler tracking, Mr. Smith employs an asset-tracking program for his shipments using a GPS-based locating device, discreetly attached to the inventory itself. In conjunction with his inventory management system, any shipments arriving now without all of the original contents, can be reconciled and the missing contents located by their GPS signals. Immediately after the implementation of his new system, he was able to identify the distribution center where inventory was being stolen and launched an investigation, leading to arrests.

Scenario 2–Natural disaster impact on crisis response

Vernon Smith has each of his distribution centers, manufacturing facilities, and offices logged and being monitored in his ARM system, which notifies him when something of significance that could impact his operations comes within 10 miles of his facilities. On one day in particular, a series of severe thunderstorms with the chance for tornadoes and flash floods were forecasted, particularly in a pattern that had the potential to impact two of his managed facility assets. Understanding that during the period expected when the storms might hit these locations that it was during business hours, Mr. Smith had to ensure that proper crisis response and preparedness procedures were being enacted, so that if the facility were hit, employees and visitors would be in a safe location to shelter in place.

This situation could also directly impact travel depending on how visiting employees or contractors react to the situation, and possibly call upon the company's crisis response hotline for advice and support services.

Scenario 3–Civil unrest impact on facilities and productivity

Mr. Smith receives a security alert from his intelligence provider that civil unrest has broken out in a city where he has a major call center. He quickly assesses the situation, looking at where the employees who work there live in relation to the facility and whether their route to work will be safe and will avoid the conflict? By understanding the proximity of the situation to his facility, Mr. Smith is able to arrange for safe transport of those employees who live close to the location without experiencing the civil unrest, while employees who would be forced to travel in close proximity to the violence were set up to work from home via remote call forwarding. Understanding where your people are in relation to your facilities in a crisis such as a natural disaster can help to avoid a major interruption in service. If you consider employees to be assets (i.e., human capital), you must treat them with a high sense of value to your business continuity plan.

This situation can also directly impact travel to and from a location, triggering temporary bans or added security protocols.

ORM is the structured ability to mitigate and manage a business's resiliency around operations in the face of interruptions. Operational risk, according to the Basel Committee, is the risk of loss resulting from inadequate or failed internal processes, people and system, or from external events.[2]

---

[2] Wikipedia, "Basel Committee on Banking Supervision," http://en.wikipedia.org/wiki/Basel_Committee_on_Banking_Supervision.

Like ERM, ORM can be employed to many more aspects of a company's operation than are listed below, but for discussion purposes, let's consider just physical security and operational resiliency.

Physical security

Examples of physical security include entry and exit points for a facility; in the context of travel, this could be the entry and exit points of a hotel. Not having good security around entrances and exits of hotels, could lead to travelers feeling unsafe and therefore negatively impact the hotel's business and "operational resiliency" per se.

Some core components of physical security are:

- *Access permissions and restrictions*—tightly managed access keys or cards, with updates or redistribution upon departure of previously authorized employees.
- *Security systems*—connected to local authorities, the notification of the policy or fire departments can be critical for mitigating the incident.
- *Monitoring*—recording who entered and left the facilities, establishing a visitor manifest and a timeline.
- *Communications*—training for employees on how to communicate different types of emergencies to key company stakeholders or the authorities, even under duress.

Operational resiliency

One example of operational resiliency, again using a hotel as an example, is to consider the risks associated with not cross-training your employees to work both the front desk and guest services. If two of your front desk attendees call in sick at the same time and you are expecting a group to check in that day, would it be worthwhile to pull in trained employees from guest services to help check everyone in? The decision to take that risk in the event that you are without the resources you need, is a small example of how operationally resilient a department or a company can be, by simply looking ahead and calculating the potential risks, along with a plan for how to mitigate those risks. In short, operational resiliency is the how well an organization can rebound or adapt to unexpected circumstances with minimal or no impact on operations.

Interestingly enough, other documented approaches to ERM such as the Corporate Treasurers Council's "CTC Guide to Enterprise Risk Management Beyond Theory: Practitioner Perspectives on ERM,"[3] are very similar to the TRM3 model, which focuses on TRM, as is noted in the Council's IAMGOLD case study, which identified the feature company's four-phase approach to ERM as consisting of:

- Risk identification and assessment
- Risk mitigation
- Risk policy
- Risk infrastructure

---

[3] Nilly, Essaides, "Enterprise Risk Management Beyond Theory: Practitioner Perspectives on ERM," Corporate Treasurers Council and Association for Financial Professionals, Inc., 2013, http://www.pwc.com/us/en/risk-management/assets/beyond-theory.pdf.

Case Study 2: IAMGOLD Corporation

This mid-size Canadian gold-mining company has a deeply rooted risk culture, which it recently formalized into a four-step process. It treats ERM as a living/breathing process as the company continues to refine its approach, and views successful risk management as a competitive advantage.

The mining business is inherently risky. It involves large capital investment, significant operating commitments, and costly exploration programs in countries that may suffer political and social instability. It's no surprise then that IAMGOLD professes to have had a strong ERM program long before it made revisions to its policies and procedures in 2012, according to the ERM team comprised of senior executives: EVP and CFO Carol Banducci; Aun Ali Khokhawala, Director of Internal Audit and Risk Management; SVP of Corporate Affairs Benjamin Little; and Treasurer Alberto Nunez. The ERM team discussed the company's program during an interview in May 2013.

"In the mining sector, there's already a heightened awareness [of] how risk can impact operations and local communities," said one executive. "It's embedded into our culture. Every time we look at our business plan and strategy, we go through a risk assessment," the executive explained. "Mining is risky and it is important for the business to understand the nature of those risks and how to deal with them."

According to these senior leaders, "ERM is not a one-time program, it's a process. There's always been a form of ERM displayed in the way the business is managed. A year ago we put more clarity around the framework about how to assess, quantify measure, and report risks." However, according to the risk-management team, while the risk culture has been prevalent, there has certainly been more recent emphasis from the financial community and the board to instill more rigor around it. "Having the process more formalized helps with the communication with the directors and the investment community," according to one senior executive.

For companies in the mining industry such as IAMGOLD, risk management is not only a necessity; it can be a powerful competitive advantage. "If you do not have that supportive culture and an excellent program you are at a competitive disadvantage," one participant said.

The four-phase process
The team said the ERM program is something the company takes seriously and is fundamental to how the business is managed. The program has four phases:

1. **Risk identification and assessment.** Define the risk universe with input from across the organization. Risks are assessed within a two-dimensional model of impact and likelihood broken into four broad categories: strategic, operational, financial, and compliance, with an accompanying structure of accountability both at the corporate level and at the various sites.
2. **Risk mitigation and reporting.** Define rules, responsibilities, control activities, and processes to mitigate and monitor those risks.

3. **Risk policy.** Document the risk policy and processes, including reporting and communications, and how ERM is integrated into the business planning process.
4. **Risk infrastructure.** Document the company's appetite for risk and implement technology tools to track the risks that impact the business and strategic plan. The company is currently at this stage.

The ERM framework was initially designed by Internal Audit/Risk Management (IARM). "IARM supports management in reporting to the board and Audit Committee [about] how we are doing versus our risk framework," the executives said. "We sit down with the Executive Leadership Team and review risks in terms of both a short- and long-term horizon and in relation to our business and strategic plans." That overview is then captured within key areas including compliance, financial, strategic and operational risks.

"We get input from all functional and site leadership," reported one executive. "We do functional, site management and executive-level reviews, and based on the collective input, we come up with the most significant risks to us."

The IAMGOLD team noted that ERM is an important, comprehensive, and proactive undertaking that is used to assess and manage the company's key risks. "It's an evolving program. Wherever there is a potential risk, we identify it, address it, and update our risk universe," they said, adding that while the key risks will get the most attention, all risks are continuously on the radar screen.

IARM is the ERM process owner in terms of developing, monitoring, and reporting protocols and their respective action items. Execution is handled by the functional and business unit executives. Specific risks are assigned to specific individuals. IARM pulls that together and reports to the board through the oversight of the Audit Committee on a quarterly basis, and more frequently if necessary.

Such board and management level buy-in is critical to the running of the program. "The engagement from those levels is absolutely necessary. You're setting the tone at the top," said one executive. "The time, effort, and rigor at the top cascade through the rest of the organization. If there is not buy-in from top management, it becomes a corporate or compliance exercise. This is not the case here. The CEO is visibly engaged and spends a considerable amount of time on ERM, supported by the board and the chair of the Audit Committee." In fact, risk management is defined as one of everyone's key objectives, which is critical to creating a culture of accountability.

ERM in practice

The risks the company identifies through its process are integrated into the highest level of management decisions, as well as day-to-day operations at the site level. "We look at risks to the business and the strategic plan. We identify mitigating activities for any risk that might prevent us from achieving those plans," the executives explained. "We go through this level of rigor at the project level. It gives us insight into risk management not just at the corporate level. ERM goes into every aspect of the business including managing our balance sheet and capital structure," they said.

"Risk management plays a significant role in the work we do with communities and governments where we operate," the ERM team said. The relationships the company builds with various local constituents help stabilize its presence and avoid potential dangers. The company refers to its work with the governments and communities as its social license to operate in that country or region. "It drives a detailed framework that involves all political elements and stakeholders," the executives said.

To highlight how critical this area is to the business, IAMGOLD designed a very specific risk management framework to help address risks related to the government dimension of the business. "It is a very robust program that affects the operational and strategic plans and ties to compliance," the executives said. "We have to operate within the legal mining framework within these countries."

The corporate affairs effort is best viewed as a subset of ERM. What sets it apart is "the degree of systemization and disciplined implementation," one participant explained. There's a process by which risks in each jurisdiction are identified, and an active program is put in place to mitigate each one. The program is run out of the corporate affairs office in Canada, but managed jointly with the country leads. It's effective because of "the significant amount of time that our most senior people spend engaged with governments in host jurisdictions," he said. "The program captures various facets of government risk, from elections, when you want to avoid becoming politicized, to a high degree of engagement with local media, opposition and incumbents, to communicating the total contribution that we make to the economy."

Added one executive: "I have seen it work negatively at companies that do not have that level of engagement. We identify periods when risk is elevated, for example, budget times." In each jurisdiction, the company identifies risks and sources of leverage, which are very specific to the location. "You want to be able to look at influences on outcomes, and make sure you understand how it works."

Commodity price exposure was identified as a key risk in the industry. The company runs through scenario analyses based on different price assumptions and establishes appropriate alternate action plans for each scenario. "In our industry the price of gold is not something we are able to control," they said. "We must have well-developed plans to adjust our business and operational plans and, if needed, we must be prepared to implement those plans. Price impacts can be material, so you have to think this through ahead of time."

Evolving program

This year, IARM is working to create a more robust, detailed risk policy and document the company's risk appetite and tolerance level. That does not mean there is not one now. The work the team did defining the process already gave rise to a substantial amount of documentation. "There is a common definition and clarity around how things are defined," explained one executive. "It is important to ensure that we define risks in a consistent way so there is a constructive conversation. There is a lot of work that has already been done in standardizing the nomenclature." The team added: "We have a lot of information and insight about the process, risk impact and the policy. As we discuss it, we continue to refine it."

Advice for others

The IAMGOLD executives offered this collective advice to their peers, in terms of key ERM success factors:

1. **Buy-in from the top.** "This is the key item for success that [we] would suggest for any ERM program. Without it, it is going to be a much less effective process. It is important to understand how this can add value to the business, from an investment, compliance, and operational perspective."
2. **A robust process.** Next in line is having a rigorous process. According to these ERM pros, the company's four-phase approach lends structure to the process.
3. **Keep it fresh.** "Don't do it once and put it away," they advised. "It needs to be a living, breathing process as your business develops."
4. **Get the right people in the room.** Finally, according to these ERM veterans, it is important to get the right people to assess the right set of risk areas.

Source: Nilly, Essaides, *Enterprise Risk Management Beyond Theory: Practitioner Perspectives on ERM*, "Case Study 2: IAMGOLD Corporation," Corporate Treasurers Council and Association for Financial Professionals, Inc., 2013, pp. 16–18, http://www.pwc.com/us/en/risk-management/assets/beyond-theory.pdf.

As the IAMGOLD case study infers, you can have as many policies and procedures as you want, but if you don't have a culture that respects the value of mitigating and managing risk from the top down, your efforts won't be effective. However, you don't need to be in an industry with seemingly high risks such as mining or oil and gas to adopt such a culture and infrastructure. As companies conduct more global business and expand in developing nations around the world, it doesn't matter if you are in software or agriculture, the need for a cultural adaptation to risk, down to the employee level, not just senior leadership, increases year over year. In the IAMGOLD case study, the "culture of accountability" instills the responsibility of risk management be shared by everyone, and with the evolution of health and safety or duty of care laws around the world, prosecution of negligence is no longer limited to senior level executives. In Australia, their PCBU (person conducting business or undertaking) expands liability and responsibility to lower-level employees involved in activities that pose any potential risks to the safety of employees.

Travel management companies 10

To set the tone in this chapter, the first thing that I would like to say is that there are vast differences between travel management companies (TMCs). It's easy to assume that they are all alike, when comparing fundamental transaction fulfillment services (reservations and ticketing), but an employer's choice of TMC to book, ticket, and report and advise from a procurement perspective, will have a huge impact on the success or failure of the employer's travel risk management (TRM) program. While any support of TRM programs by TMCs should be rigidly defined and limited so as to not have the TMC assume risk or liability for the employer's responsibility of managing risks, at the core of the TMCs value to TRM is the reservations data that drives traveler tracking. Delivering timely, standardized PNR (passenger name record) or reservations data to a global data warehouse for export to a third party, or directly to a third party for use in a risk management platform is by itself a monumental task. It is a task that gets more complicated as the number of countries and TMCs involved in the program in question increases, whether or not policy compliance for booking travel and capturing reservations data is mandatory.

For those who do not have a travel industry background, it is easy to make sweeping assumptions about how things should work with regards to how travel data is created, updated, and manipulated, how it is transferred, security considerations, etc. However, unless travel buyers have had travel agency operations experience, it is difficult to understand the level of the effort that it takes to implement a TRM solution with a TMC, and what to expect or not expect from reservations data, which is inherently inconsistent and difficult to work with on an ongoing basis. This chapter documents some of the intricate details of what travel buyers need to understand when dealing with a TMC and TRM.

Global distribution systems

There are seven primary, global distribution systems (GDSs) used around the world by TMCs, including Sabre, Abacus (a Sabre company), Amadeus, and Apollo, Galileo, and Worldspan (all three, Travelport solutions) and, lastly, TravelSky (China government owned). These systems and variations of these systems (used in different ways by TMCs) are also used by airlines for reservations, inventory/yield management, and ticketing. These systems facilitate the sale of travel-related inventory by travel agents (TMCs), which is available via the GDS, while managing inventory, revenue, and yield management (at variable levels) for the suppliers who provide access to their inventory in the GDS, such as airlines, hotels, car rental companies, and other suppliers. In some cases, the GDS simply acts as an interface to a supplier's own or third-party inventory management system. In addition to the purchase of the transportation or accommodation in the GDS, pricing tools for airline fare tariffs are provided, as well as invoicing capabilities that interface with back-office accounting and reporting tools exist. Depending upon the level of GDS participation by the supplier in question, a host of additional auxiliary products and services can be sold as well, such as special seating on the airplane, class of service upgrades, excess baggage, and more.

GDSs are often the backbone automation of the inventory sold by most Internet/ online travel agencies, as well as TMCs. Many Internet travel agencies or TMCs supplement GDS data by connecting the GDS content with "direct connect" access to a supplier's own inventory management system, serving up the combined content and availability that seems seamless in a user interface, such as with an online booking tool, whether it be for consumers or business travelers. Custom pricing and supplier inventory can be managed in the GDS, as well as travel booked outside of the GDS, which gets entered into the GDS as something called a "passive segment," We discuss passive segments in the section below (see "Passive Segments").

Computer reservations systems

Technically, GDSs are also computer reservations systems (CRSs), but this term is sometimes used for the smaller, regional systems or country-specific systems, often whose primary creation and use was prompted by a specific airline. TMCs local to the airline's market may also use these regional or country-specific systems in addition to a GDS, even when that same airline's inventory and availability are also listed in the GDS. This is done for many reasons, such as airlines only publishing some special pricing in their native CRS, or the airline charges less per booking to use its CRS. Each time a supplier's inventory is booked via a GDS, the supplier usually pays a fee to the GDS for handling. Examples of CRSs include Axess, Radixx, KIU, Mercator, and Navitaire (an Amadeus company).

Passive segments

A passive segment is manually entered reservations information for a flight, car rental, or hotel into a GDS- or CRS-based PNR that was booked outside of the GDS or CRS in question. Not uncommonly, and for various reasons, such as booking a hotel via phone in order to secure a meeting or special rate not available online or in the GDS, reservations are made outside of the GDS (sometimes in a CRS). However, if this travel is not booked via the employer's managed travel program's TMC, it can be considered "leakage" or an "open booking," yet it typically still needs to be captured into a PNR within a GDS for transfer and use in traveler tracking systems. Is it possible to get this data into traveler tracking systems without utilizing a GDS? Yes, but the most common method for capturing non–GDS-booked reservations data is to create passive segments with a GDS PNR that gets processed with all other GDS reservations data. It is not considered leakage when the TMC of record must book a reservation this way, because it is happening within the context of the corporate travel program. TMCs enter passive segments into PNRs when they make these offline bookings, and can do the same for bookings made by travelers, but that information is rarely provided to TMC agents.

Capturing these details in the GDS is a common practice so that the transport or accommodation data can be handed off to the appropriate data warehouse or directly to a TRM solutions provider. Passive segments are differentiated from active segments (booked via the GDS or CRS) using special status codes that denote that the booking was not reserved via the GDS automated inventory. Even within the same GDS, there are wide varieties of ways that a TMC can format the passive segments. Even though they ultimately serve the same purpose, there are variations in formatting that cause reservations data standardization issues that all TMCs face when trying to support global TRM programs for multinational clients who use reservations data from multiple countries and/or TMCs. Including key data elements and passive segment format specifics in service-level agreements or other forms of TMC contracts will only help to ensure that any applications requiring reservations data be less difficult to implement and manage. Some TMCs may already have global standards in place, but few do, especially in smaller, remote countries around the world.

Important note: No matter what TRM technology that a TMC offers (third party or proprietary), if it cannot deliver in real time, or in close to real time, standardized reservations data to your TRM solution in a way that captures all GDS and non-GDS transportation details and traveler's information (e.g., name, e-mail address, and mobile phone number) with a high level of accuracy and an ability to monitor and correct reservations' data rejects, employers should have serious apprehensions about using that TMC. If the TMC uses a third party to standardize variable data formats before providing it to the client's TRM solution, clients must consider the PII (personally identifiable information) implications for such a process.

When using open booking–related applications, the end result of the data output is sometimes the creation of passive segments in a GDS PNR, or directly passing itinerary data into a database that can accept and populate the relevant database, which is normally populated by GDS PNR reservations data. The problem with using open booking is the inability to ensure that all open bookings are being captured or forwarded to the applications that then convert the emails into usable data that can be fed into traveler tracking tools or other applications. Many consumer-based websites do not allow temporary holds on travel inventory, so by the time an open booking's itinerary details have been forwarded to a parsing application for database documentation, the purchase has already been made and pretrip mitigation may be inadequate.

Reservations data is the most recent version of a PNR (GDS reservation that can contain multiple transport and accommodation segments) as a result of any new segment creation, change, or cancellation. When reservations data is used with TRM systems or third-party tools for things such as pretrip approval tools, reservations data updates need to be pushed to their appropriate application databases in real time or with minimal latency.

Invoice data is not and should not be used for the population of traveler tracking or many third-party TRM applications. Invoice data contains the end result of one or more transaction's segment and financial details. Depending upon the TMC, its processes and available technology, invoice data isn't readily available until after it has been processed with payment settlement, and finalized in a back-office and/or accounting system, which can be in intervals of days or weeks, or even settled for reporting and reconciliation on a monthly basis. The frequency of delivery, and the absence of opportunity to mitigate risk and provide disclosures and training and the earliest possible time requires as close to real time reservations data only, which cannot be achieved with invoice data.

Incorrect assumption

"I understand that my TMC uses the same GDS in each of our countries where they service us, so all of the data should be structured and look the same?" Wrong. Not only can passive segments be entered differently into PNRs of the same GDS, but there are variable formats for the following items as well, such as:

1. E-mail address (specifically the traveler's e-mail address, being able to format and recognize the traveler's e-mail address from their arranger's or anyone else's).
2. Mobile phone number (specifically the traveler's mobile phone number, being able to format and recognize the traveler's mobile phone number from their arranger's or anyone else's).
3. Department codes
4. Cost Center codes
5. Cost savings codes
6. Project codes

Now imagine all of the potential format variables with one TMC and 15 servicing countries for "Client ABC" all using the same GDS A. What if five of those countries used GDS B and some of those countries are with different TMCs that are not affiliated? What you are beginning to see is one of the biggest challenges that multinational travel programs face when trying to satisfy local or regional requirements for booking in the preferred GDS for their respective region.

Educating buyers regarding reservations data (expectations versus reality)

Reservations data is inherently imperfect and within the context of a managed travel program, takes considerable time to cultivate to a high level of accuracy, which may never reach 100 percent given today's reservations data environments.

Reservations Data Case Study

Company ABC (a fictional company) awards its account to a new global TMC in a consolidated program, covering 10 countries (for example). While timelines may vary between TMCs for how long they take to implement their program, based upon the number of resources involved and how efficient their operations and account management teams are in the respective 10 countries, a new implementation with this many countries can take anywhere from 90 days to 6 months. Notwithstanding custom requirements and full TMC operational implementation plans, the following tasks relative to reservations data, must be completed by the TMC in order to facilitate the use of reservations data for TRM purposes (this is not a universal list, and it will vary by client and TMC):

1. Account creation with client reporting IDs assigned (financial and reporting systems by country and globally).
2. Human resource (HR) data feed set up for traveler profile creation and maintenance in GDS and online booking tools. This step is critically important for ensuring the accuracy of formatting and updates to e-mail address and mobile phone information in profiles, in conjunction with a profile management tool that synchronizes an online booking tool and GDS profiles with a central profile database. (Check with your TRM solutions provider whether mobile phone numbers need specific formatting, such as inclusion of international dialing codes.)
3. Creation of company and traveler profiles in the GDS and online booking tools.
4. Configuration of agent scripts for client-specific reservations processes by country.
5. Configuration of online booking tools (process, policies, preferred suppliers, etc.).
6. GDS security settings configuration for transfer of reservations data to data processor, which feeds data to the TMC global data warehouse or third-party TRM solution.
7. Configuration of automated queuing (transfer) process for each reservation to be sent to system or data processor for loading into TMC global data warehouse or third party TRM solution provider upon each reservation creation, change and cancelation. In absence of automation, incorporating agent-mandated operational scripting to queue/transfer the reservation manually is recommended, but not ideal.
8. Set up and configuration of mid-office, quality control systems. (Sometimes used for queuing or reservations data to data warehouses or directly to TRM solutions providers.)
9. Programming of traveler e-mail address and mobile phone formats per country, per GDS environment with systems or data processors, which convert raw reservations data to XML for use in TRM solutions where applicable.
10. Testing of connectivity between each country's GDS environment and the system or data processor to ensure that each reservation is passing security protocols, and e-mail address and mobile phone format programming are being captured and reported properly. For example, will reservations made through one country's after hours service, follow the same processes and formats as when booked through the normal reservations office, during normal business hours?

The previous list leaves out a multitude of other implementation plan items such as communication plans, launch plans for "go live" by country, and training, but was called out to provide context to show how much work has to go into configuring and testing reservations data (not even what is required for invoice data), before the data created can be of any use in a TMC-based or third-party TRM solution. Once everything has been set up properly, and mechanisms are in place for ongoing data quality maintenance and support, buyers/companies need to understand the inherent issues with reservations data that prevent it from ever being 100 percent perfect for use in traveler tracking or TRM solutions.

Buyers need to have realistic expectations about what any TRM solution can provide with regard to what can be reported on because of various factors impacting the reservation. The following list includes some of the factors that contribute to reservations data inaccuracy and just need to be accepted as limitations associated with using it:

1. *Incomplete GDS supplier data*–Example: Hotels that list their inventory in GDS reservations system for sale, may not always list complete address information in their listing; therefore, when an agent books one of these properties in the GDS and the data flows through into a global reporting data warehouse or TRM solution, an incomplete or inaccurate street address would fail to register a longitude or latitude, which is required for the mapping of the hotel location on a traveler tracking map.

2. *Incomplete passive segment detail*–Many passive segment formats in GDS reservations systems allow for considerable "free text" areas for the agent to include additional information about the sold segment in question. If standards are not used for the order and use of each of these sections within the passive segment, reporting technology cannot consistently tell the difference between the line that should be the street address versus the line that should be the city and country, or postal codes and phone numbers. Often these free text fields can include confirmation numbers or anything that the agent wants to include within character limitations.

3. *Supplier restrictions on passive segments*–With some airfares only available via airline websites (web fares), for this information to end up in global reporting tools or TRM solutions, passive air segments need to be created in the GDS (unless custom solutions have been created to bypass this process). However, some airlines receive messages via the GDS when a TMC creates a passive segment, and prohibits the use of them, sending back a UC (unconfirmed) status or similar message, not allowing the use of such segments in the GDS, with the potential to cancel the original reservation! In these circumstances, TMCs find creative ways to post passive segments that data processors can translate into the required data from segment formats that do not generate messages to the airlines in question. TMCs need to carefully understand and manage passive segment limitations by supplier, understanding how to capture and report the information. There can be costs to suppliers associated with a TMC using passive segments, which is one reason why some suppliers restrict their use. Many supplier inventories are not included in major GDSs because of the costs to support GDS inventory, which is a reflection of the value of a good TMC to ensure that all bookings required by clients can be made no matter where they must be booked, but also centrally reported upon, both from a reservations data and invoice data perspective. One example of a TMC "work around" for passive segments with suppliers, who restrict them, is to *not* use the airline's two letter code (e.g., AA for American Airlines or UA for United Airlines, which are used here for discussion purposes only), and replacing this code with a generic two letter code such as "YY." This may allow the creation of a generic passive segment, but doesn't adequately translate into TRM solutions. Therefore, TRM solutions have difficulty identifying any travelers impacted by a flight on a specific airline that may have been involved in an accident if the YY code was used instead of the actual carrier code. The invoice may reflect the free text name of the airline, but may be insufficient for reporting purposes.

4. *Supplier reservation takeover*–Example: A TMC makes an airline reservation and issues an electronic ticket for a traveler, who arrives at the airport only to find that his or her flight has been canceled, and the airline has taken the liberty of booking them on the next available flight. Sometimes that can be on the same airline, sometimes it can be on another airline. However, when this happens, such changes are happening outside of the managed TMC environment and may not be captured in reporting or

TRM solutions. There are exceptions to this use case, depending upon the airline and the GDS system used, but in general, such an incident is not tracked.
5. *Emergency travel centers (after hours support)*–If the TMC in question hasn't ensured that all GDS environments used by their afterhours support centers are connected to processes and automation that support reservations data reporting for global reporting or TRM solutions (i.e., GDS security settings for the emergency travel center's GDS environment to be authorized to queue/transfer reservations to the data processor), then any new booking or change, won't be tracked.

Different uses for reservations data

Pretrip approval applications

Because of how reservations data is created and reported, the need for reflecting the most recent version of a reservation and all of its components is best captured from GDS PNR data, which is the most useful in the facilitation of a pretrip approval process. Such GDS PNR data (reservations data), can be obtained in real time (or close to it) via multiple methods from most GDS systems. Unlike reservations invoice data, reservations data shows a "snapshot" of the current status of the reservations, perhaps after many revisions, whereas invoice data is the final itinerary and financial settlement data, which is often settled anywhere from weekly to monthly in reporting systems (depending upon the TMC). Many pretrip approval systems can build approval hierarchies, based upon HR data feeds, along with different process workflows for seeking approvals for trips booked within a managed program, such as a sequential approval order, or "shotgun" approval, which sends approval requests out to several potential approvers, and the first one to approve the request advances the document onto its next step in the process before a ticket can be issued. Some of these applications incorporate rules engines with the capability of supporting complex "if/then" statements for rules such as "if the ticket is over US$1000.00, then require two levels of approval." Other systems do not, and simply put each reservation pushed through their system (either via manual agent queuing processes or by sending all bookings through queuing automation), which follows a standardized approval process for all reservations. In the near future, from the work that is currently being done in the market, there is expected to be pretrip approval applications that incorporate risk-rating APIs (application programming interfaces), which will rate each destination included in the itinerary so that each approval can take into consideration not just the cost of the trip, but also the relevant risk and all applicable travel and risk policies.

Quality control systems

Another tool in the arsenal of a competitive TMC is quality control technology, which can be programmed on a per-client basis to support custom rules such as the following:

1. Date continuity
2. Proper corporate discount validation

3. Proper documentation (i.e., employee ID codes, department codes, cost savings codes, etc.)
4. Proper forms of payment validation
5. Proper contact information formatting (e-mail addresses and mobile phone numbers)
6. Specific suppliers utilization
7. Spending threshold monitoring
8. Various other types of rules that each system can support based upon searchable criteria within the PNR

In addition to helping TMCs with formatting consistency, these systems can also be the conduit for proper transmission of PNR data creation, updates, and cancelations to third parties, such as TRM providers or pretrip approval application providers. The problem is that even large, multinational TMCs rarely have these systems in place in all of their network countries around the world. They exist in major markets, but aren't always cost effective to have in smaller countries, which then forces some of these locations to resort to manual "queuing" processes (method of transferring PNR data from one GDS environment to another) instead of having an automated process, which is not subject to human error. In a large multinational company, these inconsistencies for data delivery and consolidation are typically the "weak links" in maintaining the data integrity of any TRM program. Buyers need to understand all of the possible methods for transferring PNR data to a third party or data warehouse for export to third parties.

Some GDSs allow one automatic "queue drop" to a specified pseudo city code (PCC) or office ID (OID) (depending on the GDS, different terms are used). A PCC or OID is a work environment within the GDS with its own unique identifier, which ties any reservations made within that environment to it and the originating agent or booking tool that made the reservation, along with security protocols for access. A "queue" is like a folder within a GDS environment, where you can place PNR data for various reasons (e.g., ticketing, waitlist clearance, schedule changes). However, if the PCC or OID only allows one automatic queue drop to another PCC or OID, and that allocation has been taken, then the TMCs need to look into other alternatives, such as lines of coded formats that can move over from company or traveler profiles into the PNR/reservation, which instructs the GDS to queue drop the booking in a certain place every time an agent refreshes or "end transacts" a booking, uses programmable keys (manual process; not recommended), scripted processes that force the queue drop process, or any third-party automation that has been installed within the GDS PCC or OID or that is in use with the one allocated queue drop per PCC/OID. "End transaction" means to confirm any changes made to the reservation since the previous "end transaction." For instance, if you modify a reservation and ignore it without an "end transaction," the changes won't take and become permanent. However, scripts can be interrupted and ignored by agents, making this process subject to human error. Ideally, if the PCC or OID sits within a country that has access to a quality control software product, and all PNRs drop into that the PCC or OID used for quality control purposes, then the quality control system can manage the distribution of PNR data to multiple places. However, each rule or level of complexity programmed into these quality control systems often comes with a price, which increases your total cost per transaction, but should be considered as an appropriate cost of doing business in support of TRM programs, if even available in the countries in question.

With so many variables around GDS-booked segments, passive segments, variable mobile phone and e-mail address formats, etc. within a managed TMC environment, why would anyone choose to believe that data collection for risk management purposes would be any better via open booking applications? Hopefully, the examples provided within this text provide clarity for when open booking applications show some, but limited, values with assistance in capturing leakage, or travel booked outside of a managed program. For example, an open booking process through an e-mail parsing of itinerary data, may capture the booking data into a standardized format that populates passive segments in a GDS or TRM solution. But what if that booking has a flight cancelation or schedule change and the adjusted itinerary isn't sent via e-mail to the itinerary parser to update the tracking database? As previously mentioned, even when open booking applications can assist with capturing booking travel program leakage, it still faces inherent challenges. Ensuring that any bookings made outside of the TMC environment are properly passed on to the itinerary data parser is subject to human error, with the exception of a select few major suppliers who have direct connectivity to these open booking applications. In the absence of allowing email inbox scanning, can you be absolutely sure that every time a booking is made on a consumer-based website, that the travelers will do their part to ensure that the company receives the data properly for security purposes? I think not.

Travel management company core offerings

In approximately 2011, the concept of a baseline, TMC core offering for TRM products and services was created as a standard offering for all clients, at no additional charge. The intent of these offerings was to provide valuable tools and best-in-class intelligence from a third-party risk intelligence provider, at a minimal level that would constitute a firm starting point for a client to build and create its own TRM program. The core offering was never designed or intended to be enough to address all Travel Risk Management Maturity Model (TRM3) key process areas. When the initial core offering was established by a leading, global TMC, others quickly followed suit. The offerings from each TMC varied widely. Some skimped on what they called "intelligence and alerts" from providers with little or no security analysts on staff, only employees or contractors scraping news releases and redistributing them. However, a select few did incorporate best-in-class intelligence as part of their core offering, from one of a handful of companies who employ large staffs of dedicated security analysts who have lived and worked in their respective areas and have local contacts with not just the media, but local government, police, community, and tribal leaders (where applicable), and often report on incidents happening well before the information hits the news wire services.

In the beginning, these core offerings included primarily risk-rated traveler tracking and reporting dashboards.

Risk-rated traveler tracking and reporting dashboards

Alert subscriptions for select contacts within client organizations (not travelers, as pushing content to travelers was a cost item), such as travel and security managers are typically part of these core offerings.

No communications capabilities or automatic push of risk reports or alerts to travelers were included in core offerings because these were aspects of premium solutions that required upgrades at a cost. Some of the technology provided in the core offerings was proprietary, and some were from third parties, depending upon the TMC in question. It was easy to be blinded or confused by which offerings were the best ones, because some of the traveler tracking dashboards offering long lists of reporting capabilities, but many of which may not have been of practical use to the client, or incorporated subpar (non-intelligence based), third-party content. Understanding the difference between news and intelligence is a critical part of measuring the value of any core offering or premium solution employed by a TMC client.

News is something that anyone can get online, for free from a multitude of sources, usually after an incident has already taken place. As previously mentioned, intelligence is in effect "analysis + news + context + advice," created by qualified security analysts.

More recent travel management company "travel risk management core offerings"

As TMCs try to use TRM as a competitive edge to winning more business, they continue to add products and features to their core offerings. Because each company's approach to TRM varies, no matter how lucrative a core offering becomes, they were never designed to adequately meet all of the needs of a company's TRM program. By design, they were created to be an effective means for upselling to a more advanced, paid version of the platform that does address more of the proactive notifications and disclosures mentioned in the TRM3 model. Therefore, companies shouldn't exclusively use the core offerings for long, before graduating to customized programs of their own, which usually include the custom tracking reports and notifications, automatic distribution of security intelligence (risk reports and alerts) to travelers, crisis hotlines that take into consideration different insurance providers from around the world and different policies and protocols for handling different types of incidents, based upon the company's instructions, ultimately addressing all of the TRM3 key process areas. A recent addition to TMC core offerings is the ability to provide GPS coordinates for traveler tracking via mobile applications, which feed into the traveler tracking reporting dashboards offered. These traveler locations are in addition to the itinerary-based data provided through the TMC booking process or open booking leakage data capture. Many companies and countries express concerns over these types of GPS feature functionality for privacy reasons, but many of these options only transmit a specific location via the mobile application users pressing a button on their mobile device to transmit their location for just that moment of using the feature. They do not typically provide any more details than that. However, technology exists to actually monitor all movement of a traveler, but such functionality typically involves special satellite phones, with additional costs for the technology and phones required.

Beyond TMC-provided TRM solutions, most TMCs will support a buyer's direct third-party solution, whereby the client directs the TMC to hand off reservations data to the third party who then provides the technology and intelligence, fueled by the TMC reservations data. Advantages of using third-party solutions like this are that they are typically highly customizable, and sometimes more mature solutions, and can incorporate reservations data from many different TMCs around the world. TMC-based proprietary solutions most often only support that specific TMC's data, and does not incorporate other TMC data for consolidated reporting. TMCs typically do not like to release their data to other TMCs, which is why doing so to a third party hasn't seemed to be an issue for them in the past. It's not impossible for TMC-based solutions to consolidate data from other TMCs, but it's just rare to see it. It's another good reason to try and have globally consolidated travel programs with a single TMC where possible, but it is recognized that this may not always be practical for some programs. Direct buyer–supplier relationships for third-party TRM platforms typically work better with companies who have multiple TMCs globally that need to have their data consolidated into one solution.

Standalone technology masquerading as travel risk management, versus comprehensive travel risk management

To the uneducated buyer, understanding the difference between basic traveler tracking with risk communications and comprehensive TRM can be difficult. TRM isn't something that you can "check the box" for with an off-the-shelf product, or technology alone. As mentioned in Chapter 2, TRM is a discipline that includes technology, quality intelligence, services, training, comprehensive preparedness, and response plans, all mentioned in the 10 key process areas of the TRM3 model, which is discussed in Chapter 2.

However, because risk is an ever-emerging topic of importance to companies, some technology providers of various core competencies have attempted to take a share of the market potential, and present only part of the solution as an add-on to their primary solutions. For those buyers of the mindset where they think that TRM is something that they can just "check the box" on, a false sense of coverage or completion may be experienced, but make no mistake: simply having traveler tracking and the ability to push disclosures and communicate with travelers before, during, or after a trip isn't sufficient by industry standards to meet the duty of care requirements.

With risk communications and basic traveler tracking solutions, questionable solutions that claim to be TRM solutions, consider the following questions:

1. What is the quality of the intelligence used? Is it written by qualified security analysts or scraped from news wires or the Internet and nonsecurity professionals?
2. Does the solution provider grasp the concept of TRM3, or are they simply trying to sell you an add on to their core products or solutions?
3. Can the solutions offered support expanded ERM (enterprise risk management) solutions?

4. Are crisis response support services available as part of the TRM solution being provided, and who is behind crisis response case management? Is it a known, reputable provider for both medical and security-related incidents?
5. Does the solution provide access to security analysts or subject matter experts who can provide consultation to clients on specific events or intelligence?

Finding the money for travel risk management

11

In Chapter 1, we have cited case law and country-specific legislation that mentions requirements for employers with regard to duty of care, but how does one "bring that down to a personal level that gets a board of directors' attention?"

Hydro One's Qualitative Risk Appetite Rating Scale

hydro one

Three Powerful Questions
- Only three questions are needed to get a good sense of executives' risk appetite.
- Asked in sequence, these questions help register a real understanding of committing to a risk appetite.

Rating	Risk Taking Philosophy	Tolerance for Uncertainty How willing are you to accept uncertain outcomes?	Choice When faced with multiple options, how willing are you to select an option that puts this objective at risk?	Trade-Off How willing are you to trade-off this objective against achivement of other objectives?
5-Open	Will take justified risks	Fully anticipated	Will choose option with highest return, accept possibility of failure	Willing
4-Flexible	Will take strongly justified risks	Expect some	Will choose to put at risk, but will manage the impact	Willing under certain conditions
3-Cautious	Preference for safe delivery	Limited	Will accept if limited, and heavily out weighted by benefits	Prefer to avoid
2-Minimalist	Extermely conservative	Low	Will accept only if essential, and limited possibility/extent of failure	With extreme reluctance
1-Averse	"Sacred" risk avoidance is a core objective	Extremely Low	Will select the lowest risk option, always	Never

Example questions for understanding your organization's risk threshold or "appetite."
Source: Provided by the Association for Financial Professionals, Inc. Nilly Essaides, "Enterprise Risk Management Beyond Theory: Practitioner Perspectives on ERM," Corporate Treasurers Council and Association for Financial Professionals, Inc., 2013, p. 25, http://www. pwc.com/us/en/risk-management/assets/beyond-theory.pdf. Copyright 2013, Association for Financial Professionals, Inc.

At a high level, before presenting a Travel Risk Management Maturity Model (TRM3) business case to board members or executives in support of investing in programs and solutions to help with global travel risk management, lays the foundation for the conversation with the following context:

1. Who is responsible for duty of care?
    a. For someone to justify a claim in common law negligence, they must first prove that a duty of care was owed them. In general, courts test this question with the following criteria:
        i. Is there a relationship between the parties involved?
        ii. Was the damage or injury foreseeable?
        iii. Is it reasonable, fair, and just to impose a duty of care?
2. Was the duty of care breached?
    a. When duty of care has been determined to be established between the parties, the claimant must be able to prove a breach took place.
        i. This is where proof of an industry standard, and how the party with the duty of care lives up to such a standard, is critical (i.e., a TRM3 analysis)
            1. A TRM3 analysis can provide proof of some standard and comparison against other companies, of what duties or efforts are reasonable, as well as if the company is meeting those standards according to the analysis (separate from the specific incident at hand) over time if the TRM3 assessment is a part of an annual continual process improvement exercise.
3. Were there damages?
4. Were the damages in question foreseeable or preventable?

In the absence of a formal travel risk management (TRM) program, most companies will struggle to answer these questions sufficiently to defend themselves against a claim of duty-of-care negligence if sued.

The value of a human life

Finally, we get to the unpleasant topic of how some companies or governments come to associate a financial value with a human life. Based upon previous topics covered in this book, the potential for employer losses can vary widely, potentially resulting in a collectively large amount, depending upon the type of incident, visibility of the incident, and the impact on business continuity. However, one factor that sometimes comes into play within some corporate cultures when assessing what to spend on certain mitigation strategies or response and recovery missions is how one measures the value of the life, or remainder of life, in question. Economists consider the value of a statistical life (VSL) when people consider the risk-to-reward tradeoffs with regard to their health. However, this is different from the value of an actuarial life, based upon the likelihood of death. In an actuarial life, you have to consider the average life expectancy of the person in question based upon demographics, as well as how much of that person's life has statistically lapsed or taken place, based upon their age.

In 2015, the U.S. Transportation Department noted the statistical value of a single life as US$9.4 million,[1] but there are many models used for different reasons.

---

[1] U.S. Department of Transportation, "Guidance on Treatment of the Economic Value of a Statistical Life (VSL) in U.S. Department of Transportation analyses — 2015 Adjustment Memorandum, June 17, 2015, https://www.transportation.gov/sites/dot.gov/files/docs/VSL2015_0.pdf.

While most people believe that a human life is priceless, some companies do look at variable statistical values of a human life to determine some risk thresholds and what they will and will not spend in terms of mitigation or potential for damages.

Baseline considerations for your TRM business case could include:

1. Build your case with relatable examples, using your companies top travel destinations or processes impacting travel (possibly meetings and events). Use the following examples for inspiration to find your own:

   a. Do you have five or more employees traveling to the same destination at the same time? Can you detect any flight with an exceeded number of employees traveling on the same flight, before they travel? Remember the company with 20 employees on Malaysian Airlines flight 370? Imagine trying to defend the company for not being able to prevent so many employees from traveling together at once? What kind of costs in damages were incurred, as well as loss of business continuity and reputation?

   b. Do you mandate a travel policy that includes exclusive use of company-designated travel management companies (TMCs)? If the company doesn't have pretrip insight into reservations data, including where and when people are traveling, there is little potential for mitigating risks prior to travel, providing adequate training and disclosures to travelers.

   c. Does your company have internal teams, along with third-party crisis response resources to have planned protocols and responses to support medical or security related emergencies including, but not limited to:

      i. Ebola
      ii. Civil unrest
      iii. Earthquakes
      iv. Tsunamis
      v. Hurricanes/typhoons
      vi. Individual traveler emergencies (sickness, assault, etc.)
      vii. Death of a traveler on the road

   d. How prepared is your company should a traveler get hurt at a hotel that the company required use of, in the absence of safety and security standards for corporate-preferred hotels?

2. Provide examples of international law that reflect both financial penalties and criminal penalties for company executives, and in some countries (e.g., Australia), for mid-level managers as well who are found negligent in their duty of care.

3. Provide each investment stakeholder with the iJET whitepaper on the TRM3 risk assessment process for context (regardless of what TRM solution or supplier you use).[2]

4. Provide each investment stakeholder with the benchmarked TRM3 analysis scorecard for your organization, showing how your company rates in each key process area. Most companies beginning this process score almost completely reactive and unprepared for fulfilling what would be considered "reasonable best efforts" compared to industry peers for duty of care standards. This is why complete honesty is essential in conducting the TRM3 analysis, so that plans can be employed for real and tangible improvements that result in positive change and reduced risk exposure.

---

[2] iJET, "Travel Risk Management & Maturity Model (TRM3)," http://info.ijet.com/resources/whitepaper.

5. Compile a list of all insurance and contracted medical and security providers for your company globally, and conduct a cost-to-benefit analysis for the following:
   a. Will your current insurance providers discount your premiums with the implementation of a comprehensive TRM program?
   b. Is it beneficial to put your global insurance requirements out to bid for cost savings and global standardization of standards and services.
      i. Most TRM solutions providers are partnered with global medical and security insurance companies who can provide competitive solutions at competitive rates in conjunction with a TRM program.
6. Share the cost of parts or all of the program with departments or customers:
   a. Transactionalize the cost of your TRM solution via your TMC and include it in your TMC transaction fee. If possible, include these fees with your billable expenses to clients for project work.
   b. Find funding from departments other than travel to supplement or cover the cost of TRM solutions (Legal, Human Resources, Security, HSE (health, safety, environmental), etc.).

The following is an example of how to break down your annual TRM solution subscription or license fee costs into your TMC transaction fees:

Company A uses ABC123 Travel Management Company (fictional) globally, which supports approximately 10,000 transactions annually at a cost of $30 per transaction (for discussion purposes only). The annual cost of Company A's TRM solution license fees are estimated US$30,000, excluding case fees or third-party services (e.g., includes technology, intelligence and crisis hotline). While Company A may be required to prepay the cost of the US$30,000, it can perhaps recoup most or all of its TRM program subscription or license fee costs if its travel costs are billable to its customers in the following manner:

Annual TRM base costs:	US$30,000
Divided by number of TMC transactions:	10,000 transactions
Equals cost per transaction for base TRM program:	US$3.00 per transaction
Plus the standard TMC transaction fee:	US$30.00 per transaction
Equals combined TMC/TRM fee per transaction	US$33.00 per transaction

Example baseline components of TRM subscription based solutions available for licensing include:

1. *Basic itinerary-based traveler tracking*—Cost item with many premium, TRM solution providers. Free with some TMC core offerings.
2. *GPS-based traveler tracking*—Cost item with many TRM solution providers. Free with some TMC core offerings. This feature transmits a latitude and longitude based location in the TRM solution reporting dashboard for the moment when the traveler transmits a "check in" signal. It does not provide tracking of continuous movement.
3. *Intelligence-based travel alerts*—Cost item when distributed to travelers, using quality risk intelligence (not news) with either TMC or TRM solution provider. Some TMCs provide no-cost alert subscriptions to a small number of individuals at each of client's companies that are involved with TRM.

4. *Messaging capabilities*—Cost item with both TMC and TRM solution provider programs. Often includes traveler communications from system reports via e-mail, SMS text message, or push message via mobile applications.

5. *Crisis response hotline*—Sold by TMCs or TRM solution providers, but always powered by TRM solution providers, this hotline should be one consolidated resource/phone number for travelers to call from anywhere in the world should travelers experience a crisis, such as medical or security issues, theft, assault, or intellectual property loss. This is separate from a TMC's after-hours emergency hotline for reservations.

6. *Special intelligence reports*—Airline safety reports, hotel safety assessments, health assessments based upon geography, etc.

7. *Satellite tracking*—With the use of satellite phones and special TRM solution features, select user's movements can be tracked as the user moves. This is especially important for some industries working in remote, isolated areas, such as offshore oil rigs or remote, unpopulated areas. This is a cost item, typically only provided by TRM solutions providers.

8. *Global medical services and evacuation network access only*—An "access only" program is typically for those companies who have insurance in place that will pay for these services as needed, but do not necessarily have the global resources to support the client in question. The "access provider" can provide the support and can coordinate payment with the client's insurance provider(s).

9. *Global medical services and evacuation network access plus insurance*—In the absence of separate corporate insurance relationships to pay for access and services via a global medical services and evacuations provider, both the access to the network and the insurance to pay for services rendered would be provided by the same supplier. Most major TRM solutions providers have partnerships to offer both "access only" and "access plus insurance" options. It is important that your TRM solution provider, who does the case management via its crisis response hotline for any medical or security-related services, be agnostic as to who the client chooses to use for medical services or payment, even if it has a preferred partner in this area. This is because many clients have preexisting relationships that they want to maintain, which can and should be managed within the full scope of the client's TRM and crisis response protocols with their TRM solution provider.

How do I show a return on investment on risk management?

In essence what you are trying to achieve is to show value for things that don't happen, because in fact you are managing the risks well. If you use the fundamental concept of "How would we be able to defend ourselves against litigation, should we need to prove that we made 'reasonable best efforts' as compared to our peers within our industry?", then detailed, year over year documentation of those efforts is a good place to start.

That happens with following the TRM3 assessment process as part of an annual business process improvement exercise with benchmarked analysis reports. Investment will be required to raise your scores on the assessment to acceptable levels and then to maintain them on an ongoing basis. The potential for loss, which should also be a part of the return on investment (ROI) discussion, should call investor's attention to:

1. Loss of reputation.
2. Loss of operations or business continuity.
3. Loss of trust from employees and customers.

4. Loss of ability to obtain and keep new talent (risk management and safety are key considerations for new hires in today's global economy).
5. Loss of cash or capital for lack of preparation or coverage. Always consider the cost for the transfer of financial risk via insurance products, or a cost benefit analysis for self-insurance covering:
   a. Medical-related incidents
   b. Security-related incidents
   c. Fleet-related incidents
      i. Liability coverage
      ii. Personal accident coverage
   d. Business interruption insurance
      i. Stop loss limitations
      ii. Liability coverage
      iii. Excess liability coverage
      iv. Political risk/liability coverage

Savings via self-insurance

Simply doing the initial and ongoing analysis for whether or not to self-insure, can be costly and time consuming, but for large organizations whose employees travel internationally, self-insurance is another effective tool as part of an ERM (enterprise risk management) strategy. However, self-insurance, in essence, is based upon a company or organization allocating enough money to compensate for potential losses, based upon predictable and measurable risks using claims histories (usually spanning a minimum of 2 years) and legal/regulatory requirements for minimum coverage depending upon the item, person, or operation being covered, along with the jurisdiction. Self-insurance doesn't necessarily mean no insurance, as is the case with providing legal minimums for fleet-related liability insurance via self-insurance, along with excess liability insurance where additional coverage may be needed by market or jurisdiction.

Many companies with large vehicle fleets under management self-insure damage to the vehicles themselves, but have some sort of personal accident and liability coverage largely because estimating these amounts can often be difficult, in contrast a predetermined value of the vehicle assets.

While self insured companies can sometimes benefit from costs savings depending upon the findings of their claims reviews and financial analysis, savings must be balanced with the added cost of policy administration and claims processing, which can sometimes be outsourced to third party administrators.

Additional benefits of being self-insured can include customized plans, catering to the specific needs of the business, and faster claims processing and enhanced cash flow.

Business continuity or operational resiliency plans

If you could, sit down with your CTO and ask the question, "What is the cost to our organization for one hour of network downtime?" When it comes to critical infrastructure and operations, many IT departments know these numbers as they are a part of business cases for disaster recovery, or business continuity plans. According to Gartner[3], based upon industry surveys, the average cost per minute of network downtime is $5,600 USD, which is over $300k USD per hour. In contrast, think about the potential costs or losses associated with project delays, lost sales, manufacturing inefficiencies relative to your inability to safely move key personnel in and out of work or client site locations, where their presence is required? According to the U.S. Travel Association, statistical models over 18 years and 14 industries indicate that for every dollar invested in business travel, U.S. Companies have experienced a $9.50 return in terms of revenue[4]. With that in mind, it is arguable that the resources and support provided through a managed travel program and TRM, be looked at holistically as investments in business continuity or operational resilience, and perhaps partially funded from those budgets accordingly.

Using OSHA (Occupational Safety & Health Administration) – U.S. Department of Labor Data for TRM Investment Business Case

In the United States, workplace safety guidelines are enforced at the federal level by OSHA (Occupational Safety & Health Administration). Business travel is widely agreed upon as an extension of the workplace, from a legal and operational perspective; therefore, while unofficial, the case could be made that statistics on savings relative to workplace safety inspections and programs via OSHA for example, investment in TRM programs and resources should be considered in kind. On the OSHA.gov website[5], OSHA cites a study showing a 9.4% drop in injury claims and a 26% average savings on workers' compensation costs during four years after a Cal/OSHA inspection in comparison to a similar group of uninspected work locations.

Leveraging ERM (enterprise risk management) identifying claims and costs savings opportunities

Building a business case for investment in TRM can be very challenging, especially if your foundation is primarily based upon cost avoidance. One of your biggest challenges may be doing the due diligence necessary to identify, document and correlate potential cost savings to the company, based upon your proposed TRM program. This is where there may be some opportunity for consolidation between TRM and your

---

[3] http://blogs.gartner.com/andrew-lerner/2014/07/16/the-cost-of-downtime/.
[4] https://www.ustravel.org/sites/default/files/Media%20Root/5.2015_BizTravel_Report.pdf.
[5] https://www.osha.gov/dcsp/products/topics/businesscase/.

overall ERM (enterprise risk management) strategy; whereby combining some areas of risk mitigation could generate some incremental savings and justify any potential newly proposed expense for TRM.

Consider the following:

- Insurance review – global review and analysis of medical, security and evacuation coverage globally, to ascertain if consolidation or joint offering with TRM provider can provide significant cost savings.
- Outsourcing and consolidating IT risk or loss related support requests to your risk management supplier (e.g., lost or stolen laptop, mobile phone, etc.) (utilizing the same supplier used for TRM).
- Is your company tracking hard and soft costs associated with business travel interruptions? If so, what expenses were incurred over a minimum of a two-year period? Are there any opportunities there for process improvement and savings?
- Is your company tracking specifically any medical or security related expenses for workers traveling abroad? Are there any efficiencies to be gained from these incidents had they been managed under your TRM program?
- How much time, and what was the relative value of that time, was spent over the last year to two years by internal resources managing travel security or medical cases and claims management for business travelers?
- Location/site threat assessments—combining this aspect of ARM (asset risk management) with your ERM (enterprise risk management) strategy, which includes TRM (travel risk management), may provide added cost savings benefits to your business case, if consolidated with the same risk management supplier.
- Executive decision support—intelligence based analysis of risks associated with doing business in a particular market or area, or with particular groups or businesses. Where has your company paid for expert advice relative to business related risks, and could those be leveraged via ERM.

TRM – setting new standards for what's to come

12

Travel risk management (TRM) strategies can and often do change regularly, especially within the travel management company (TMC) community.

Since 9/11, and with growing intensity and interest year over year since then, the topic of duty of care and proactive TRM continues to be one of the most talked about among travel managers/buyers and the business travel community. Although more than a decade has passed since 9/11, and with all of the discussion, webinars, and reading material on the topic, many TMCs and buyers are still great lengths away from managing and understanding TRM at the level that they should. Some TMCs and some non-risk, technology solution providers often make a lot of noise about how great their programs are, but if you look "underneath the hood" at what they have in place, their programs often don't go far beyond traveler tracking and maybe some alert distribution and communications, casting aside many of the Travel Risk Management Maturity Model (TRM3) key process areas required to bring TRM programs up to industry parity for "reasonable best efforts." Misinformation about what is and isn't "enough" or "reasonable best efforts" can easily be skewed when big brand names in the travel industry attempt to offer solutions and claim to be subject matter experts on risk, when nothing could be further from the truth. Choose your solution providers and your partners wisely, based upon their holistic knowledge of TRM and how well TRM is understood throughout each TMCs organization in detail (not lip service).

When looking for a TMC or a TRM solution provider for products, services, and intelligence to support your program, buyers should consider asking the following questions:

1. Who is responsible for developing and evolving your TRM product strategy? What is their background? How long have they been working with a focus in TRM?
2. To what extent is TRM a part of the supplier's core competency or business model? In other words, what percentage of the supplier's annual product development and training budget is spent to ensure that its solutions are competitive and that its employees can speak to those solutions with complete confidence and in great detail?
3. Can any of your TMC or TRM solution sales or account management people articulate the importance of TRM to their customers, and what makes your solution the most competitive in the market? (Many companies now providing TRM solutions spend initial funds on development and marketing, but then fail to execute launch plans that include internal training. This shows a lack of commitment to the subject matter, and a lack of expertise.)
4. Can you provide a list of updates and new features that you have made to your solution over the last 2 years? Some companies have acquired technology or partner with third parties and have not made updates in months or even years.

5. Is your solution proprietary or third party? If third party, have you invested in custom development of the product to make it your own? If so, to what extent?

6. Who provides your risk intelligence content? Describe the kind and amounts of resources dedicated to creating and maintaining your travel risk intelligence? Also, what is the vetting process prior to distribution?

7. Does your solution and risk intelligence partner (if applicable) include crisis response call/case management support? Is this services supplier agnostic to suppliers they will coordinate with for clients (e.g., medical services)?

8. Does the solution provider in question offer TRM solutions as checkboxes on a list of other products and services offered or is there a consultative approach offered when addressing your TRM needs?

9. Can their solution or their intelligence partner's solution, support a graduation from TRM to support enterprise risk management (ERM)?

Travel management companies

The vast majority of TMCs have one or more partnership relationships with TRM solution providers. These relationships are most commonly referral relationships, whereby sales leads are simply passed to their partner for an incentive once a contract is signed directly between the supplier and the buyer. However, there are circumstances where the TMCs act as resellers and can provide TRM solution provider products (and sometimes custom, partner involved solutions) on their contract paper as a reseller, if the client requests it, streamlining billing and paperwork. There are a few TMCs that license intelligence content or technology to include in their own white-labeled, third-party solutions, and even fewer who license the intelligence for use in their own proprietary technology.

A travel management company's role in travel risk management

TMCs talk a lot about TRM, and it is a topic that gets covered in almost every sales presentation or proposal. Some TMCs even have designated personnel who are responsible for supporting such programs, whether they are proprietary or third party, but the fact remains that from a cultural acceptance perspective, adopting the importance of TRM into their business model to the point of their people fully understanding it and being able to teach best practices, is only happening with a small number of TMC thought leaders. Just as airlines have developed incremental new revenue streams to their business models for ancillary services such as premium seat assignments or baggage fees, TMCs are conceivably in a prime position to assist customers with developing and implementing programs, solutions, and best practices, considering they can learn from the variables experienced through dealing with their diverse customer bases globally. However, even some of the world's largest TMCs historically have struggled to execute a consistent, recommended TRM approach or strategy and typically have very few sales executives or account managers who can speak with confidence about TRM. With a few exceptions, what typically happens is that either the sales or account manager contacts the one person (possibly two) within the global organization who is a supposed subject matter expert, to have that person support sales calls or client consultations on the topic of TRM, or sales or account manager contacts whichever third-party TRM solutions provider partner the company has to ask for some support, so managers don't have to learn the subject matter in detail themselves. Why is this so important? It's important because when we talk about TRM for companies, we are talking about something that impacts the safety of the people who traditionally travel for your company, along with all of the other potential losses that we have discussed throughout this book, including how TRM touches enterprise-wide risk. Employers need to make sure that anytime they solicit advice on TRM, that the person from whom they are getting advice truly knows the subject matter well and has extensive experience in the area. What you don't want is someone who just repeats talking points from brochures or proposals. As TMC's start to embrace TRM as a more standard part of their business model, as a few thought leaders are starting to do, TMC account managers will hopefully be ideally situated to more comfortable speak to TRM at a basic level in conjunction with a subject matter expert, or as more commonplace with their TRM solution partners in joint efforts.

Important areas where TMCs could play valuable future roles in customer TRM are:

- *TRM assessments*—The first and critical step in understanding where each client is in their TRM program development, how each client compares to other similar companies, and what each client needs to get to industry parity.
- *Sourcing*—After assisting with assessments, if the client needs support for sourcing an appropriate TRM solution (and supporting services), a TMC partner should help with these exercises, just as they do for other spend categories relative to travel, such as airlines, cars, and hotels. Because TRM programs are such intimate reflections of a company's culture across multiple departments, such a sourcing exercise will still need close interaction with client stakeholders and cannot be conducted solely by the TMC with data alone.
- *TRM platform solution implementations*—Even if a client uses multiple TMCs globally, because of the technical and operational expertise required to connect and automate reservations data delivery, a senior TMC account manager or technical project manager should take the lead for project management of a new TRM solution implementation, unless the customer has this experience. Along with assigned TRM solution implementation resources, clients need a primary stakeholder to oversee the set up documentation and delivery of the following at a minimum:
  - TMC operations for the client in question in the country(s) required, that support the TRM solution's data requirements, using globally standardized traveler profiles where possible.
  - GDS (global distribution system) security settings and branch access (or alternate delivery methods) in order to transmit reservations data in a timely fashion across borders.
  - Client identifiers (account numbers) and additional required data identifier formats (employee ID, department or project codes, if required, and traveler e-mail addresses and mobile numbers) in the reservations data, and document how they reside in the data and are provided to whatever data processor service provider that initially receives and

converts the data into an XML or required format for the TRM platform solutions provider, or to the TRM platform solutions provider itself.

- Identification and implementation of an automated method for transmitting each reservation upon every "end transaction" (creation, change, or cancelation) to the contracted data processor or TRM platform solutions provider (avoiding manual queuing/transmission of passenger name records [PNRs] if at all possible).
- Tests to ensure that the data is not only being successfully received and processed into the TRM platform solution, but that it contains all of the necessary, required data included in the TRM platform solutions provider's implementation documentation.
- Establish senior support contacts across all participating TMCs in a global program matrix, should any collaboration be necessary, or simply for the benefit of the buyer/client for use during and after implementation.

- *TRM platform solution maintenance*—If a client is using a TRM platform solution solely for traveler tracking and not for communications, disclosures, and risk mitigation, they simply aren't meeting minimum thresholds for "reasonable best efforts" when it comes to duty of care. Case in point: because TRM is a discipline and ongoing program that needs maintenance and support even to maintain basic tracking, it is not a turnkey solution that one can implement and walk away from once the client has tracking capabilities. In addition to TRM3 key process areas, for an effective TRM program to remain effective, TMCs could conceivable help with strict attention to:
    - *Report on PNR leakage (open bookings)*—Reservations not captured in the TRM solution, even if an open booking application is used. In some instances, the supplier may have direct connectivity for travelers who book with the supplier outside of the supplier's managed program, but in many instances, as mentioned earlier in this text, such bookings rely on travelers emailing their itineraries for processing into the TRM solution or having their email inbox scanned for itineraries. Reporting on those bookings, even if it is post trip, can help with identification of problem behavior and compliance. However, the entire goal for TRM is to be able to mitigate risk before it happens, thus one of the major drawbacks from allowing open booking.
    - *Report on PNR rejects*—TMCs, TRM platform solution providers, or data processing services must have the means to produce reports for those PNRs that are missing key data points required for the client's TRM platform solution, such as:
        - Traveler e-mail address (not the booker's or administrator's)
        - Traveler mobile phone number (in the proper, specified format, typically with country codes)
        - Client IDs or account numbers
        - Employee IDs (optional)
        - Department numbers (optional)
        - Project codes (optional)

Being able to quickly identify these PNRs as rejects, and to promptly have TMC operations follow up on the missing data elements for reprocessing so that the system works properly, is a critical quality control process that clients must insist be supported. TMCs should actively monitor the following:

- Advance notice of any operational changes impacting TRM PNR data flow and capture, such as the following, should be provided to clients and TRM solution providers by TMC's, in order to prevent any gaps in data or solution performance problems:
    - TMC GDS conversions
    - Implementation of new or changing GDS pseudo city codes or office IDs.

- Change in technology or methods for automatic queuing or delivery of PNR data to the data processing service or TRM platform solution provider (example.g., going from a GDS automatic queue drop set up to using a mid-office, quality control solution for queuing based upon programmed routines).
- Any changes to reservations data-formatting structures.

Data quality and travel risk management program maintenance

Important things to note regarding reservations data and TRM solution implementation and maintenance are:

1. *Consolidated TMC reservations data feeds (where applicable)*—TMCs that provide a single global data feed (e.g., XML api) to TRM solution providers are preferable because the time required to implement a country is less with this delivery method, versus GDS system direct connectivity (queuing) per country is required.
2. *Data quality*—Even a consolidated global data feed from a TMC doesn't ensure that the data being provided is completely usable. Within the one data feed, there can still be formatting inconsistencies or missing data from country to country, such as traveler e-mail addresses (or erroneously placing a travel arranger's e-mail address in the reservation instead of the traveler's address) or mobile phone numbers, as well as variable formatting for passive segments. Passive segments need consistency in formatting so that the TRM solution provider knows how to read them.
3. *Project requirements*—An employer/client's involvement in defining data requirements, impacting agency operations, human resource (HR) data feeds, and any third-party application programming interfaces (APIs), is critical in setting and managing expectations around how well a TMC can support the solution requirements. Employers should minimally address the following requirements:
   a. HR data feed
      i. Method of data delivery to TMC (for traveler profiles) and TRM solution provider (custom data requirements not in profiles)
      ii. Frequency of data delivery
      iii. Provide HR data schema
         1. Required data for use in TRM solution
         2. Required data for use in TMC profiles
            a. Identify where profiles are used and where they are not
   b. Required data for use in TRM solutions
      i. Client/account ID
      ii. Country identifier (GDS and pseudo city code or office ID)
      iii. Traveler name
      iv. Traveler e-mail address (specifically the traveler's, and not someone else's unless TRM solutions provider supports supplemental e-mail addresses, in addition to the traveler's)
      v. Traveler mobile phone number *with* international dialing codes standardized. It is absolutely critical that mobile phone numbers be standardized in an agreed upon, global format across all countries, in order for SMS communications to work properly. Some TRM solutions providers can implement "workarounds" in the

absence of a global formatting standard if there is at least a consistent format on a per-country basis, but this only makes the implementation and maintenance of good data quality that much more complex

    vi. Active segment reservations data (air, car, hotel, rail booked within the GDS)

    vii. Passive segment reservations data formatting per GDS, per country (air, car, hotel, rail booked outside of the GDS)

    viii. Employee ID (optional)

    ix. Department code (optional)

    x. Other customer data fields (optional)

c. Quality control processes

    i. Identify how reservations missing required data fields, or with improperly formatted data (PNR rejects) get identified and reported

    ii. Identify process and service-level agreement timelines for how quickly PNR rejects get corrected and resubmitted into updated reservations data

    iii. If possible, identify a reconciliation process to see what reservations were reimbursed, but never recorded in the TRM solution

You can't manage it if you can't measure it!

With the adoption of iJET's TRM3 model as the industry standard process and framework for TRM assessments and programs, TMCs should adopt this assessment process as a standard part of their annual client review process. At the time of writing this book, very few have begun to implement this as a standard annual service for clients, and are only beginning to adopt it for select customers. In time, as TMCs hopefully become more involved as strategic stakeholders in supporting their client's TRM programs, an annual TRM3 assessment should be done for all clients, regardless of size, location, or global footprint. Unless multinational companies are consolidated with one TMC globally, it can be a challenge to use a TMC-provided TRM solution, because of the privacy and political issues surrounding other TMCs sharing data with a competitor. This is why TRM provider direct solutions can sometimes be the best path for companies operating in multiple countries around the globe with multiple TMC's—it is easier to consolidate their data across multiple TMCs. It is not impossible to consolidate data from multiple TMCs with one TMC for a tool contracted via a specific TMC, but it is not common, and not without unique difficulties.

While there are many entrants into the market for TRM solutions, not all offerings are the same. Companies cannot truly say that they offer duty of care or holistic TRM solutions unless they can provide support for each of the key process areas of the TRM3 model. Therefore, suppliers should fall into one of two categories: technology partners or providers (those that offer some automation relative to TRM and may partner with a full TRM solution provider for intelligence and sometimes customized an/or white labeled solutions, but may not cover the TRM3 key process areas [KPAs] themselves) and TRM solution providers (those that offer support for each of the TRM3 KPAs and usually produce their own risk intelligence and provide a host of other risk management support services and solutions).

What the future holds for travel risk management

Without a doubt, one of the biggest struggles with regard to effectively implementing and maintaining a TRM program is quality reservations data. It's not enough that you receive and can read the individual segments on an itinerary (air, car, hotel). You must be able to properly capture the traveler's e-mail address and mobile phone number so that important risk disclosures, as well as any crisis or ad hoc communication that the employer is trying to distribute, can be communicated. As we have discussed in earlier chapters, travel data (specifically reservations data) is inherently "dirty" or not standardized, making it difficult to work with. Even if you are lucky enough to scrape, parse, or otherwise capture a mobile phone number or e-mail addresses, are you sure that the data belongs to the traveler and not to the booker or arranger? It is those kinds of details that can derail the effectiveness of your program, because if you don't have it right and the traveler never receives the disclosures in question and suffers damages that could have been prevented had they received the pertinent information, the employer will likely be held responsible.

For many years, the travel agency or TMC community has worked toward the concept of the "universal PNR" or "perfect PNR," which theoretically would standardize all scoped data points within the PNR from segment detail and traveler contact information, to custom reporting fields and cost savings data, but the reality is that it still doesn't exist industry-wide. However, some global TMCs have made considerable progress toward achieving this goal, while others remain far, far behind.

Alternative means for capturing sufficient traveler data

Corporate card data

Corporate credit card preauthorization data has been integrated and used as an alternate source of travel location detail, combined with merchant detail provided post sale, such as airline flight itinerary details. However, travelers must use integrated corporate cards that would provide such data. So what about those travelers who do not have corporate cards? There are some questions around how quickly the itinerary detail to be carried over into the TRM solution can be obtained in traveler tracking solutions, thus calling into question the effectiveness of using this data. This capability has been developed by some corporate card issuers, which capability should be evaluated individually for the quality, timeliness, and effectiveness of the data with various TRM solutions.

Mobile application data

Let's break this down into two types of mobile applications for use with TRM:

1. *Open booking travel applications*—While these mobile applications support other features and functionality, they are widely promoted for their ability to capture travel data for travel booked by a traveler who books his or her travel with whomever the traveler chooses.

2. *TMC-hosted/promoted travel applications*—These mobile applications may be proprietary or licensed by an employer's TMC, which when used in conjunction with a consolidated and mandatory, managed travel program, should conceivably provide an improved ability to capture open bookings, TMC-generated bookings, and also serve as a conduit for communications and distribution of important disclosures to travelers.

3. *TRM solution provider mobile applications*—Depending upon the supplier, these applications can come in two categories. First, are applications that are very similar to the TMC-hosted applications, which rely on travel itinerary data. Second, are satellite applications that can monitor a traveler's every move, versus only where they are when they check in on their mobile device. The satellite based tracking applications are typically for very high-risk destinations and locations where normal mobile service is unavailable.

Conceivably, an employer could potentially use multiple types of mobile applications at the same time, but the more applications that you introduce to your traveler population, the more questions and confusion employers will encounter. In addition, your TMC may not have deployed its mobile applications in all of the countries that you require, which can also be a setback. The ideal situation for whatever travel program mobile application is used (this may vary from country to country, or by TMC) would be that your global TRM solution provider can integrate intelligence distribution (alerts and risk reports), as well as ad hoc, administrator-generated messages. These ad hoc messages could be broadcast messages to all users, or to targeted travelers from reports generated by the TRM system platform.

Mobile Application Case Study

Company ABC (fictional) uses Acme Travel (fictional) in 18 of 20 countries around the world, and policy dictates that all travel must be booked via Acme Travel where available. Policy also states that all travelers must maintain current traveler profiles in the Acme Travel portal/profile system, in parallel with Acme Travel's mobile application using the same primary e-mail address and mobile phone numbers. The goal of this case study is to optimize use of available mobile technology to populate the employer's contracted TRM solution with timely, up-to-date itinerary and contact information as quickly as possible for use on a pretrip, mid-trip, and posttrip basis.

Case study conditions:

- Two of Company ABC's 20 countries do not book their travel via Acme Travel because Acme Travel doesn't have operations in the two countries in question.
- Three of Acme Travel's 18 countries that service Company ABC, do not have the Acme mobile travel application deployed yet. This means that the application has not yet been deployed in the countries in question.
- Company ABC uses a third-party solutions provider for online booking and expense reporting, which offers an open booking mobile application to travelers for capturing travel data, which is available globally.
- Company ABC provides its TRM solution provider with an HR data feed that includes employee ID, e-mail address, and mobile phone numbers.

It's difficult enough to properly capture and report on an e-mail address, but mobile phone numbers are even more complicated because most systems that use mobile phone numbers for calls or SMS text messaging, need the exact, proper formatting of the number (including the international dialing code before the number itself) in order for the number to be successfully used. However, if each mobile application used is set up to include the user's e-mail address, matching the e-mail address listed in their profile and HR data feed, the e-mail address can be the data element or identifier that is used to match the traveler's itinerary data with the right client in the TRM solution, which correlates the traveler and the traveler's data to the appropriate HR data feed, which has the traveler's properly formatted mobile phone number. When this match is complete, not only are the itinerary data and e-mail address properly populating the TRM solution database, but the correct mobile phone number with appropriate international dialing info, can be mapped to the trip in the TRM database as well.

Such a matching process can solve the following problems:

1. It could conceivably help with issues relative to each of the 15 Acme Travel countries being used with the Acme Travel mobile application, to capture and map the reservations data to the TRM solution provider should some countries have trouble with data handoffs (or queuing) to individual suppliers. For example, a daily or hourly feed of each country's reservations data may be automatically from the mobile application or TMC global data warehouse to the TRM solution provider. The Acme Travel mobile application should also be able to capture any leakage or open bookings made by travelers in these countries, if the travelers are trained as to how to e-mail the itineraries to the proper e-mail address for capture and reporting of reservations data or if the application supports email inbox itinerary scanning. Alternatively, if all bookings are set up to copy the mobile application via API connections or e-mail with automatic itinerary e-mail distribution, the mobile application could be the vehicle for feeding TRM solutions itinerary data, including TMC booked and open booking reservations.

2. One could eliminate the problem regarding capturing of traveler mobile phone numbers and proper international dialing code formatting, via the matching process with the TRM solution provider, when using the e-mail address as the common data identifier for matching purposes.

3. In the three countries where the Acme mobile travel application isn't deployed, as long as the reservations data provided to the TRM solution provider includes the matching e-mail address identifier, the matching process can still take place and the TRM database can be properly populated with itinerary details, traveler e-mail address, and traveler mobile phone number (with international dialing codes). Transferring reservations data under this scenario to the TRM solutions provider can be achieved in a number of different ways, but Company ABC should document the method for these countries and make best efforts to ensure that the methods used are automated (versus any manual agent process) and can also support non-managed bookings (perhaps using their online booking and expense provider application if available in the appropriate language and location).

4. In the two countries where Acme Travel is not used, the open booking mobile application from the online booking and expense provider is used, and is configured to forward itinerary data with matching e-mail address identifier (in as close to real time

as possible) directly to the TRM solution provider, so the matching process can still take place and the TRM database can be properly populated with itinerary details, traveler e-mail address, and traveler mobile phone number (with international dialing codes). The use of this third-party "open booking mobile application" would have to include the promotion of policies and training about how to ensure that travelers are forwarding reservations details to this application, so that the intended processes successfully work unless email inbox scanning for reservations data is acceptable to the company.

Your TRM program is dependent upon timely, accurate reservations data, and chasing it for consistent delivery and quality, is an ongoing job that requires monitoring and resources. All buyers should take into consideration these responsibilities when negotiating the job responsibilities and costs associated with managed travel programs, which may or may not be included in the dedicated or designated TMC account manager role requirements or as specified in your TMC contracted service-level agreements.

Buyers and all key account management and TMC operational leadership should create, document, and execute a standard process for identifying PNR rejects, and how to recycle or reprocess them quickly in order to properly include them in the TRM solution. Sometimes this can be achieved simply by comparing daily reservations data processing reports between TMC's and TRM solution providers, for any anomalies.

Index

Note: Page number followed by "*b*" and "*f*" refer to boxes and figures, respectively.

Printed in Great Britain
by Amazon